Burke

ORGANIC NOMENCLATURE: A PROGRAMMED INTRODUCTION

WORKBOOK SUPPLEMENT

Prentice-Hall
Foundations of
Modern Organic Chemistry
Series

KENNETH L. RINEHART, JR., Editor

ORGANIC NOMENCLATURE: A PROGRAMMED INTRODUCTION

Third Edition

James G. Traynham

Professor of Chemistry
Louisiana State University

PRENTICE-HALL, INC., ENGLEWOOD CLIFFS, NEW JERSEY 07632

Library of Congress Cataloging in Publication Data

Traynham, James G.
Organic nomenclature.

(Prentice-Hall foundations of modern organic chemistry series. Workbook supplement)
1. Chemistry, Organic--Nomenclature--Programmed instruction. I. Title. II. Series.
QD291.T72 1985 547'.0014 84-17992
ISBN 0-13-640780-3

THIRD EDITION

ORGANIC NOMENCLATURE: A PROGRAMMED INTRODUCTION

James G. Traynham

Interior design: Theodore Pastrick
Manufacturing buyer: John Hall

Printed in the United States of America
10 9 8 7 6

ISBN 0-13-640780-3 01

PRENTICE-HALL INTERNATIONAL, INC., *London*
EDITORA PRENTICE-HALL DO BRASIL, LTDA., *Rio de Janeiro*
PRENTICE-HALL OF AUSTRALIA PTY. LIMITED, *Sydney*
PRENTICE-HALL OF CANADA, LTD., *Toronto*
PRENTICE-HALL OF INDIA PRIVATE LIMITED, *New Delhi*
PRENTICE-HALL OF JAPAN, INC., *Tokyo*
PRENTICE-HALL OF SOUTHEAST ASIA PTE. LTD., *Singapore*
WHITEHALL BOOKS LIMITED, *Wellington, New Zealand*

To

Professor C. D. Hurd,

and to the memory of the late

Professor R. K. Summerbell

—two who were models for combining
praise-winning undergraduate teaching
and praise-winning research.

CONTENTS

PREFACE TO THE SECOND EDITION

Among chemists, familiarity with a disfavored name sometimes breeds contempt for an emphasis on precise nomenclature. Some chemists who insist (correctly) on using plural verbs with *data* and *media*, or wince on hearing "between you and I," seem unconcerned, to my bewilderment, about using incorrect names for compounds about which they are reporting. Clinging to a familiar (no longer approved) style of nomenclature for personal use is understandable and maybe excusable, but teaching that style to beginning students can hardly be defended. In this edition, some of the names (and emphases) have been changed to protect the innocent.

Because proposals for modifications must be evaluated with great care by official committees to insure that a solution to one naming problem does not generate another (perhaps broader, more serious) problem, rules and preferred styles of chemical nomenclature change slowly at best. Yet during the past 14 years (since publication of the first edition of this book), at least some of the preferences have shifted, and new rules have begun to emerge. The central function of a compound name is to communicate clearly the structure of the compound. In a real sense, the name is a substitute for a structural formula. If one cannot move easily from name to formula, one either is ignorant of the nomenclature rules or is confronted with an incorrect name. This book is intended to reduce the likelihood that beginning students will encounter either embarrassment. IUPAC rules have permitted a variety of styles to reduce the burden of devising or comprehending a correct name without impairing the central function. When in doubt, the systematic name is the best choice, especially in written communications, and that style receives major emphasis in this book.

During the past several years, *Chemical Abstracts* has adopted a rigid approach to nomenclature, primarily for convenience and economy in indexing: a one compound, one name rule. Some chemists have been chafed by the fact that the *Chemical Abstracts* name is not always the one likely to be used at present in other chemical communications. The *Chemical Abstracts* name is an indexing name, however, and is not necessarily intended for general use. In this edition, attention is given to *Chemical Abstracts* names, especially when they differ from the IUPAC names that are in general use.

In a few cases (particularly in Chapter 14), the presentation has been guided more by the current recommendations of official nomenclature committees than by the currently approved rules. This choice aims at promoting the central function of nomenclature (structural information) and teaching beginning students the way the rules are likely to be.

The format of the book is unchanged, but the content has been substantially changed.

The order of classes of compounds has been changed to permit aromatic groups to be included in all the functional group classes treated, and a few polycyclic aromatic systems have been included. Official recommendations that certain names be discontinued have been followed, and whole sections of the first edition have been deleted, even when the names were ones I learned as a student and felt a certain attachment to.

To play James Lipton's venereal game* for a moment, I hope that this new edition will, even more successfully than the first edition, change for new beginners an embarrassment of names to an aplomb of names.

James G. Traynham

*Lipton, James, *An Exaltation of Larks, or, The Venereal Game*, 2nd ed.; Grossman Publishers: New York, 1977.

PREFACE TO THE THIRD EDITION

A chemical name identifies the chemical, but it can do much more. The official rules of organic chemical nomenclature have grown out of emphases beyond mere identification, emphases on conveying information about structure and about the expected chemical behavior of the compound named. Extensive investigations of organic chemical behavior, however, had been described in the chemical literature before structure for compounds, as we know it, was even acknowledged, much less known. Names, not based on structure, were devised and used.

The rapid growth of organic chemistry and the organic chemicals industry after 1850 generated intellectual excitement and challenge—and communication difficulties. When the number of chemists and compounds is small, systematic nomenclature is a nicety; when these numbers swell, it becomes a necessity. The spur toward systemization was and continues to be retrieval of information from the chemical literature. The precise use of nomenclature does not guarantee efficient retrieval, but its faulty use, particularly in indexing, virtually assumes loss of time or, in the extreme, of the information associated with that entry.

Systematic organic chemical nomenclature has emerged slowly and is still unfinished business. The first official effort was made in 1892, when thirty-four prominent chemists from nine European countries met in Geneva, Switzerland, as an International Commission for the Reform of Chemical Nomenclature. They adopted a page-and-a-half of forty-six rules (resolutions), mainly for aliphatic compounds. Prominent among these rules was the selection of the longest continuous chain of carbon atoms as the basis for a substitutive name and the advocacy of a single name for each compound. These Geneva rules gained substantial, but not complete, acceptance and usage. Some old, familiar names, not reflective of structure, persisted in usage, and new compounds presented problems not considered at the Geneva meeting.

A second international effort in organic nomenclature was initiated after World War I and led to a set of sixty-eight rules that was adopted by the International Union of Chemistry (IUC) meeting in Liège, Belgium, in 1930. A significant modification of the Geneva rules was the focus on the principal functional group for numbering of the parent chain. Another was the abandonment of a single official name for each compound; that is, flexibility in style of names was permitted.

Subsequent official development, refinements, changes, and expansions of the rules have been made through approved reports of the Commission of Nomenclature of the

International Union of Pure and Applied Chemistry (IUPAC, successor to IUC). The latest collection of the rules for nomenclature of organic chemistry is a 559-page book published in 1979. The IUPAC Commission has been stimulated and guided by proposals from individuals and chemical society committees interested in a particular area of organic chemistry as well as by the indexing practice of *Chemical Abstracts.*

The work toward completely satisfactory nomenclature rules continues and will have to do so as long as new chemical species are being discovered. Experimental work always runs ahead of good nomenclature practice, but communication about the experimental work is dependent upon comprehensible nomenclature. Generating fault-free nomenclature rules requires the same kind of care and attention to details as does generating fault-free experimental data or compelling theories. Although we do not yet have consensus on nomenclature rules for some species, familiarity with the rules that are firmly established can minimize identity problems in chemical communication. This workbook has been written (by a teacher who cares about the language we use to communicate chemistry) with the intent of facilitating that familiarity and with the conviction that each new class of organic chemistry students is the place to start developing it.

J. G. T.

ORGANIC NOMENCLATURE: A PROGRAMMED INTRODUCTION

1
Alkanes

Learning nomenclature as well as the chemical behavior of organic compounds is greatly simplified when the compounds are divided into classes. Classification depends on the types of bonds between atoms. Perhaps the simplest class of organic compounds is that of the alkanes, compounds composed solely of carbon and hydrogen with only single bonds between pairs of atoms. Alkanes may also be called hydrocarbons, a name that signals the combination of hydrogen and carbon. Hydrocarbon is a name indicating only the types of atoms present; alkane indicates not only the type of atoms but also the type of bonds that bind them together (only single bonds between pairs of atoms).

In stable organic compounds, the valence, or bonding capacity, of carbon is four, and the valence of hydrogen is one. The simplest alkane contains one carbon and four hydrogens and can be represented by the structural formula

```
    H
    |
H—C—H
    |
    H
```

An alkane containing two carbons can be represented by the structural formula

```
    H   H
    |   |
H—C—C—H
    |   |
    H   H
```

Note that both of these formulas indicate four bonds to each carbon and one bond to each hydrogen.

1. An alkane containing three carbons can be represented by the structural formula

______________________.

2. This formula indicates ______ (number) bond(s) to each carbon and ______ (number) bond(s) to each hydrogen.

3. For convenience in writing, condensed structural formulas are used most frequently. The carbons are still written separately, but hydrogens bound to each carbon are not. Condensed structural formulas for alkanes containing one, two, and three carbons, respectively, are CH_4, $CH_3{-}CH_3$, and ______________.

4. Note that the condensed structural formulas still indicate the correct valence for each atom. Counting each hydrogen as one, we find that the number of bonds indicated for each terminal carbon in $CH_3{-}CH_2{-}CH_3$ is ______ (number) and for the center carbon is ______ (number). Condensed structural formulas rather than expanded ones will nearly always be used in this book and by practicing chemists. Unless otherwise indicated, "structural formula" in this book will mean "condensed structural formula."

To name compounds we use stems that signify the number of carbon atoms present in the group of atoms being named. The stem signifying one carbon atom is meth, that for two carbon atoms is eth, that for three carbon atoms is prop (rhymes with hope), and that for four carbons is but (rhymes with cute). The stem is combined with an ending characteristic of the class of compounds. The characteristic ending for alkanes is ane.

5. The name for CH_4 is methane, formed by combining the stem ______, signifying one carbon atom, and the ending ______, indicating class of compound.

6. In similar fashion, $CH_3{-}CH_3$ is named ______, and $CH_3{-}CH_2{-}CH_3$ is named ______________.

7. A condensed structural formula for a compound named butane is ______________.

Items 8 through 20 are concerned with the formula

```
CH3—CH2—CH—CH—CH2—CH3
         |   |
        CH3 CH—CH2—CH3
             |
            CH3
```

8. Since this formula contains only carbons and hydrogens and has no multiple bonds between pairs of atoms, the class of compound it represents is ______________.

9. Complex alkanes can be named by using the longest continuous chain of carbon atoms as the basis of the name. (Note: The adjective is continuous—not horizontal or vertical or straight—but continuous.). The longest continuous chain, or parent chain, of carbon atoms in the formula above contains ________ carbon atoms.
(number)

10. Draw a continuous line through the carbon atoms in this longest continuous chain.

$$\begin{array}{llllll} CH_3-&CH_2-&CH-&CH-&CH_2-&CH_3 \\ & & | & | & & \\ & & CH_3 & CH-&CH_2-&CH_3 \\ & & & | & & \\ & & & CH_3 & & \end{array}$$

11. Stems signifying more than four carbons in the group of atoms being named are mostly Greek (a few are Latin) in origin. For example, pent signifies 5; hex signifies 6; hept, 7; oct, 8; non, 9; dec, 10; and so on. The stem that signifies the number of carbon atoms in the longest continuous chain above is ________, and the alkane name for this chain is

________________.

12. All the groups attached to the chain of carbon atoms through which the line was drawn in item 10 are called substituents. There are ________ substituents shown in the formula
(number)
above.

13. Substituents that resemble alkanes (that is, are composed of only carbon and hydrogen with only single bonds between pairs of atoms) are named by adding yl to the stem that signifies the number of carbon atoms in the substituent. For example, a one-carbon substituent, CH_3, is named methyl: the stem, ________, always signifies one carbon, and the ending, ________, signifies a point of attachment to something else. The stem eth always signifies ________ carbon atoms; a substituent with that number of carbon atoms is named ________________. The general stem for an alkane-like grouping is alk, and the general or class name for an alkane-like substituent is ________________.

14. The names of the three substituents in the formula in item 10 are ____________________, ______________________________, and ______________________________.

15. Whenever two or more of the substituents in a formula are alike, a prefix such as di (for 2) or tri (for 3) is added to the substituent name to indicate the correct multiplicity. For example, two methyl substituents will be designated not by methyl methyl, but by

____________.

Substituents are cited in a name in alphabetical order (ethyl before methyl). Only the name of the substituent is used in the alphabetical ordering; a multiplying prefix (di, tri, and so forth) that is not a part of the name of the substituent is not considered when alphabetizing.

16. Substituent names precede the parent chain name as modifiers to make a single-word substitutive name. A partial name for the formula, which specifies all substituents in alphabetical order as well as the parent chain, is ______________________.

17. A correct name, such as that given in item 16, specifies the total number of carbon atoms in the formula and may be checked easily by comparison with the formula itself. The total number of carbon atoms shown in the formula is ________ (number); therefore, the name must specify that same number of carbon atoms. The parent chain name specifies ________ (number) carbon atoms, and the substituent names specify ________ (number), and ________ (number), respectively. The name thus specifies a total of ________ (number) carbon atoms.

18. For a substitutive (IUPAC) name, the atoms in the parent chain are numbered from one end of the chain to the other. The direction of numbering is chosen so that the substituents are assigned the lower possible numbers designating positions along the parent chain. Each substituent must have a locant. The locants to be used for the substituents in this formula are ____, ____, and ____.

19. Locants are set off from each other by commas and from the letter part of the name by hyphens; each immediately precedes the particular substituent that it modifies. The complete name for the formula, then, is

____.

20. Let us reexamine the name. The portion heptane refers to ______________________

____.

The portions ethyl and dimethyl refer to ______________________ on the parent chain; there are ________ (number) substituents on this chain. The locants designate ______________________

____.

Whenever locants occur together in a name, they are separated from each other by ________; all locants are separated from the letter portions of the name by ________. The use of hyphens to separate locants from each other, or commas to separate locants from the other parts of the name is incorrect.

21. Now generate a name to go with the formula

$$
\begin{array}{ccccccccccccc}
 & & CH_3 & & & & & & CH_2-CH_3 & & & & \\
 & & | & & & & & & | & & & & \\
CH_3 & - & CH & - & CH_2 & - & CH & - & CH & - & CH_2 & - & CH_3 \\
 & & & & & & | & & & & & & \\
 & & & & CH_3 & - & CH_2 & & & & & &
\end{array}
$$

The basis for the name of the compound represented by this formula is the ______________ chain of carbon atoms. This parent contains ________ carbons. The stem signifying this number of carbon atoms in a group is ________, and the alkane name for a parent compound of this many carbons is ________. There are ________ (number) substituents on the parent chain; these substituents are named ________, ________, and ________.

22. Numbers indicate positions along the parent chain, and it is possible to number the chain from either end. One direction alone is usually correct, for the rule to be followed states that the substituents must have the smaller possible numbers. We compare the locants term by term and make the choice of set at the first point of difference. For example, 1, 2, 4 will be chosen over 1, 3, 3. Numbering from the left in the formula above assigns to the substituents the positions ____, ____, and ____. Numbering from the other end of the parent chain would assign to the substituents the numbers ____, ____, and ____. Since we must use the smaller possible numbers, the correct position designations are ____, ____, and ____. Each substituent must have a locant in the name; the locant for the methyl group is ____, and the locants for the ethyl group are ____ and ____. When two or more locants occur together in a name, they are separated from each other by __________, and locants are separated from the other parts of a name by __________.

23. For the formula in item 21, we are now ready to write a complete, substitutive name, which is ____.

24. The basis of the name for the formula

$$
\begin{array}{ccccccccc}
 & & CH_3 & & & & CH_3 & & \\
 & & | & & & & | & & \\
CH_3 & - & CH & - & CH_2 & - & C & - & CH_3 \\
 & & & & & & | & & \\
 & & & & & & CH_3 & &
\end{array}
$$

is the chain containing ________ (number) carbons. The alkane name for a parent compound of this many carbons is ______ , and the complete substitutive name for the liquid compound represented by the formula is ____.

25. The name for the liquid compound represented by the formula

$$\begin{array}{l} CH_3{-}CH_2{-}CH_2{-}CH_2{-}CH_2 \\ \qquad\qquad\qquad\qquad\qquad\quad | \\ CH_3{-}CH_2{-}CH_2{-}CH_2{-}C{-}CH_2{-}CH_3 \\ \qquad\qquad\qquad\qquad\qquad\quad | \\ \qquad\qquad\qquad\qquad\qquad\; CH_2{-}CH_3 \end{array}$$

is ________________________________.

26. A structural formula for 3-ethyl-3,4-dimethylhexane is

________________________________.

27. A structural formula for 2,4,5-trimethyloctane is

____.

ISOMERS

Compounds that have the same molecular formula (which merely indicates the number and types of atoms present in each molecule) but different structural formulas are called isomers. For example, the compounds represented by the formulas

$$CH_3{-}CH_2{-}CH_2{-}CH_3 \quad \text{and} \quad \begin{array}{c} \quad CH_3 \\ \quad | \\ CH_3{-}CH{-}CH_3 \end{array}$$

(both have the molecular formula C_4H_{10}) are isomers. The structural formulas are different: One shows a continuous chain of four carbon atoms, but the other shows a branched chain of four carbon atoms. The structural formulas represent two different compounds with different physical and chemical properties.

28. Compounds that may be represented by the structural formulas

$$CH_3{-}CH_2{-}CH_2{-}CH_2{-}CH_2{-}CH_3 \quad \text{and}$$

$$\begin{array}{c} \phantom{CH_3{-}CH_2{-}}CH_3\phantom{{-}CH_2{-}CH_3} \\ \phantom{CH_3{-}CH_2{-}}|\phantom{{-}CH_2{-}CH_3} \\ CH_3{-}CH_2{-}CH{-}CH_2{-}CH_3 \end{array}$$

have molecular formulas of ______________ and ______________, respectively, and are called ______________.

29. There are three isomers of molecular formula C_5H_{12}. The three (condensed) structural formulas are

______________________, ____________________, and ____________________.

30. In each of these three structural formulas, the valence of carbon is __________, and the valence of hydrogen is __________.

31. The substitutive names of the three isomers in item 29 are ______________, ______________________, and ______________________________, respectively.

Most alkane isomers containing fewer than seven carbon atoms may also be differentiated by use of structural prefixes, but only if there are no other substituents present. In the chemical literature, both old and current, one encounters the prefix n-, meaning normal or unbranched: for example, n-pentane. The IUPAC names of the unbranched isomers do not contain any structural prefix, however; the unmodified name means the unbranched isomer and is unambiguous. The archaic n- prefix has never had support in official rules and should never be used.

Prefixes are used for branched-chain isomers. The structural prefix iso signifies a single carbon branch at one end of the parent chain. The prefix iso is not separated from the alkane portion of the name in any way according to IUPAC rules.

32. There are two isomers with molecular formula C_4H_{10}. Butane is the IUPAC name for the isomer represented by the structural formula ______________________________________, and isobutane is the name for the isomer represented by the structural formula

______________________________________.

Note that the stem in both names (butane and isobutane) indicates the total number of carbons in the compound. Isobutane may also be named as a substituted alkane; the parent chain name is ________, and the substitutive name for the compound is ________.

33. The stem (to signify all the carbon atoms) to be used in the name for

$$\begin{array}{c} \qquad\qquad\quad CH_3 \\ \qquad\qquad\quad | \\ CH_3-CH_2-CH-CH_3 \end{array}$$

is ________, signifying a total of ________ (number) carbon atoms. The appropriate structural prefix is ________, and the complete IUPAC name using this prefix is ________.

34. The name for the compound represented by the formula

$$CH_3-CH_2-CH_2-CH_2-CH_2-CH_3$$

is ________, and the name for its isomer,

$$\begin{array}{l} CH_3-CH-CH_2-CH_2-CH_3 \\ \qquad\quad | \\ \qquad\; CH_3 \end{array}$$

is ________ (use structural prefix) or ________ (as substituted alkane).

35. The structural prefix iso is restricted to compounds with a single carbon branch at one end of the parent chain. No structural prefix is accepted to indicate the C_6H_{14} isomer with structural formula

$$\begin{array}{l} CH_3-CH_2-CH-CH_2-CH_3 \\ \qquad\qquad\quad\;\; | \\ \qquad\qquad\;\; CH_3 \end{array}$$

This isomer must be named as a substituted pentane and will be called ________.

36. Two isomers with molecular formula C_5H_{12} are pentane and isopentane, whose structural formulas may be written, respectively,

______________________________ and ______________________________.

37. A third isomer with molecular formula C_5H_{12} is known. By restricting ourselves to a valence of 4 for carbon and a valence of 1 for hydrogen, we can write only one other structural formula for C_5H_{12}. That formula is

______________________________.

The name of the C_5H_{12} isomer that contains one carbon bonded only to other carbons may be formed by use of the structural prefix neo. Like iso, neo is not separated from the alkane portion of the name.

38. The permissible IUPAC name with a structural prefix for the C_5H_{12} isomer described in item 37 is ______________________________. This prefix is not used in the specific, nonsubstitutive name for any other alkane.

39. The IUPAC name for the compound represented by the formula

$$\begin{array}{c} \quad CH_3 \\ \quad | \\ CH_3—C—CH_2—CH_3 \\ \quad | \\ \quad CH_3 \end{array}$$

is ______________________________.

Isomers with more complex branching than that signified by the prefixes iso and neo, and alkanes containing more than six carbon atoms, are not named by use of structural prefixes. Names for these compounds are based on a substituted parent chain.

40. The IUPAC name for the compound represented by the structural formula

$$\begin{array}{c} \quad CH_3 \\ \quad | \\ CH_3-CH_2-C-CH_2-CH_2-CH_3 \\ \quad | \\ \quad CH_3 \end{array}$$

is ____.

CYCLOALKANES

Cyclic hydrocarbons are named in much the same way as are acyclic ones. The operational prefix cyclo precedes the alkane name that would be used for a parent chain containing the same number of carbons as are present in the cycle or ring. Thus the formula

$$\begin{array}{lll} CH_2-CH_2 & & \\ | & & CH_2 \\ CH_2-CH_2 & & \end{array}$$

represents a liquid hydrocarbon named cyclopentane. Like iso and neo, the prefix cyclo is not separated from the alkane portion of the name in any way according to IUPAC rules. For convenience, cycloalkanes are most frequently represented by geometric figures, such as

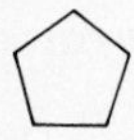

for cyclopentane. Such figures symbolize a carbon at each corner and as many hydrogens as are necessary to complete a valence of 4 for that carbon.

41. The symbol □ represents a compound whose molecular formula is ____________,

whose structural formula using Cs and Hs is

____________________, and whose name is ____________________.

42. The symbol ⬡ represents a compound named ____________________.

43. Substituents on the ring are treated just as substituents on a chain for naming purposes. Thus

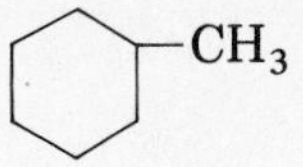

$-CH_3$

is named methylcyclohexane. The name of

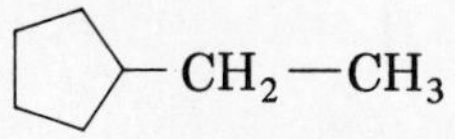

$-CH_2-CH_3$

is ______________________________.

44. All the positions in a cycloalkane ring are equivalent, and a number is not needed to indicate the position of substitution in monosubstituted rings. If there are two or more substituents, however, locants are used to indicate positions of substitution. One of the substituents is always assigned position 1, and the smaller possible numbers (locants) are used for all others. If the same set of locants would follow the selection of different substituted positions as position number 1, the preferred selection has the first-named substituent on position 1. (Remember: Substituents are named in alphabetical order.) For example, in the name of the compound

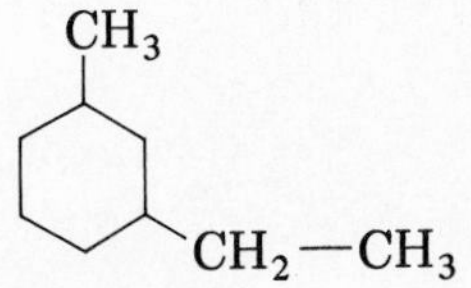

CH_3

CH_2-CH_3

the substituent to be named first is ____________, and the number of the position to which it is attached (that is, its locant) will be ______. The locant of the other substituent will be ______, and the complete name for the compound will be ____________________.

45. The name of

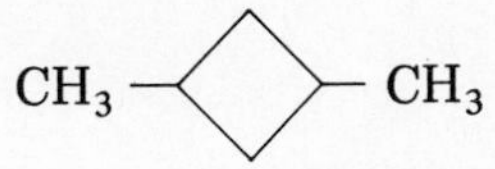

CH_3 — — CH_3

is ______________________________.

46. The use of the lowest possible set of locants always takes precedence even over the association of the first-named substituent with position 1. In the name for

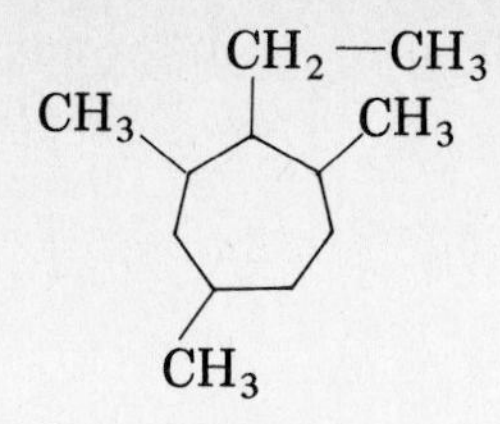

the substituent ________ is named first. If the position to which it is attached is number 1, the set of locants to be used would have to be ________. If one of the methyl groups is on position 1, however, lower sets of locants are possible. The lowest set of locants is________, and the name of the compound is ________________________________.

2 Nomenclature of Alkyl Groups

A chemical group that can be imagined to be formed by loss of a hydrogen atom from an alkane is named by replacing ane of an alkane name with yl. The general term for such a group is alkyl group.

1. When an alkyl group contains only one carbon, it is designated a ____________ group; when it contains two carbons, it is designated an ____________ group. Methyl and ethyl are definitive terms, but definitive names for higher alkyl groups must differentiate isomers.

Substitutive names are the most generally applicable and, in spite of first appearances, the simplest ones. They are used exclusively by *Chemical Abstracts* for indexing. In substitutive names of alkyl groups, the position from which the hydrogen of an alkane can be imagined to have been removed is always position number 1, and the longest continuous chain of carbons beginning with position number 1 is the parent chain. For example, the alkyl group

$$\begin{array}{c} CH_3—CH_2—\underset{|}{CH}—CH_2—CH_3 \end{array}$$

is named 1-ethylpropyl (not 1-ethyl-1-propyl and not 3-pentyl). The name 3-pentyl has about the same acceptability as the expression, "I ain't."

2. The parent chain that is the basis of the substitutive name of the alkyl group

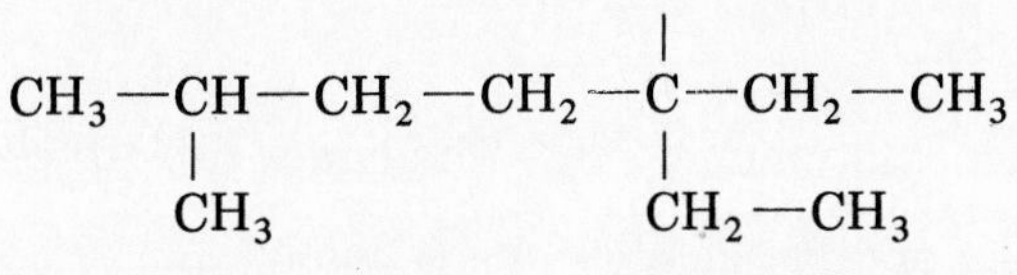

$$\begin{array}{ccccccccccccc} & & & & & & & & | & & & & \\ CH_3 & — & CH & — & CH_2 & — & CH_2 & — & C & — & CH_2 & — & CH_3 \\ & & | & & & & & & | & & & & \\ & & CH_3 & & & & & & CH_2—CH_3 & & & & \end{array}$$

contains ________ (number) carbons and ________ (number) substituents. The set of locants that must be used for these substituents is ________. The complete name of this alkyl group is ____.

3. The compound represented by the formula

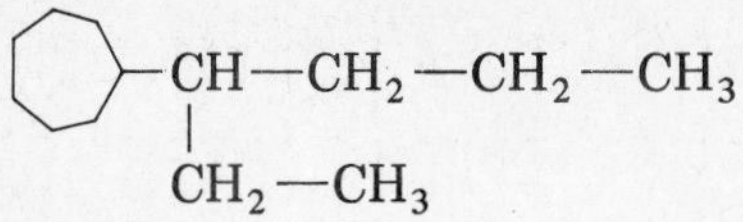

$$\text{(cycloheptyl)}-\underset{\displaystyle CH_2-CH_3}{\underset{|}{CH}}-CH_2-CH_2-CH_3$$

may be named as a substituted cycloalkane. The alkyl substituent is named ______________________ and the compound is named ______________________.

Parentheses around the name of a complex substituent may be necessary for complete clarity in a name: for example (1-ethylbutyl)cycloheptane. The parentheses make clear that ethyl and butyl are parts of the same substituent rather than separate substituents.

4. The isomeric alkyl groups

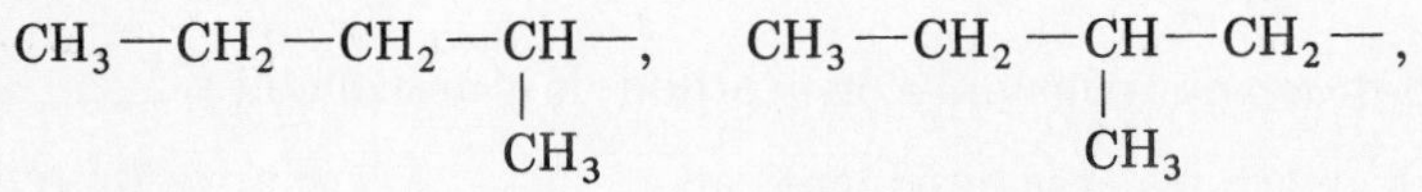

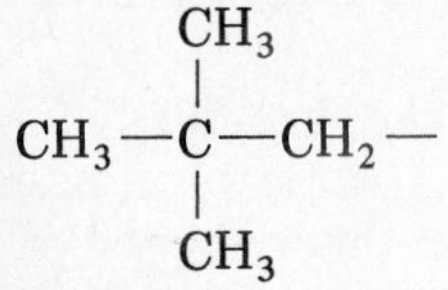

$$CH_3-CH_2-CH_2-\underset{\displaystyle CH_3}{\underset{|}{CH}}-, \quad CH_3-CH_2-\underset{\displaystyle CH_3}{\underset{|}{CH}}-CH_2-, \quad \text{and}$$

$$CH_3-\overset{\displaystyle CH_3}{\overset{|}{\underset{\displaystyle CH_3}{\underset{|}{C}}}}-CH_2-$$

have the substitutive names ______________, ______________, and ______________, respectively.

5. The formula

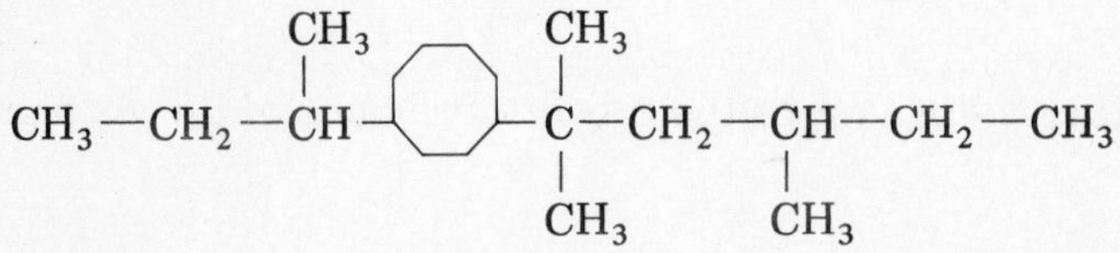

$$CH_3-CH_2-\overset{\displaystyle CH_3}{\overset{|}{CH}}-\text{(cyclooctane)}-\overset{\displaystyle CH_3}{\overset{|}{\underset{\displaystyle CH_3}{\underset{|}{C}}}}-CH_2-\underset{\displaystyle CH_3}{\underset{|}{CH}}-CH_2-CH_3$$

shows two substituents attached to a cyclooctane ring. The larger substituent is named ______________________, and the smaller one is named ______________.
The names of the two substituents appear in the complete name of the compound in alphabetical order; for complex substituents, such as those in the formula above, the initial letter of the full name of the substituent, even if it is part of a multiplying prefix, is the one used for alphabetizing (that is, d is the key letter of a substituent named 1,1-dimethylpropyl).

The first-named substituent in this compound is ______________________.

For a symmetrical parent ring (chain) such as cyclooctane, the first-named substituent is assigned the smaller position number (locant). The two substituents are on positions numbered ____ and ____. The complete name for this substituted cyclooctane is

____.

Some isomeric alkyl groups containing fewer than seven carbons may be differentiated by use of structural prefixes combined with the stem designating all the carbons in the group. These nonsubstitutive names are usually called trivial names and are also IUPAC names. In this context, "trivial" does not mean "unimportant," but it does mean "not systematic." For some of these simple groups, the trivial names have been used more frequently than have the substitutive ones, but emphasis on single names for indexing and for dependable machine searching of the literature is shifting usage in favor of the substitutive names. Trivial names are likely to be used for some time, however.

Trivial names of alkyl groups are best mastered by understanding the classification of alkyl groups. An alkyl group whose number 1 position (substitutive name numbering) is a carbon bound to only one other carbon ($CH_3—CH_2—$, for example) is classified as a primary alkyl group; one whose number 1 position is a carbon bound to two other carbons is classified as a secondary alkyl group; and one whose number 1 position is a carbon bound to three other carbons is classified as a tertiary alkyl group.

6. The stem signifying four carbons in an alkyl group is __________, and the name for the group represented by the formula $CH_3—CH_2—CH_2—CH_2—$ is ______________. This group is classified as a ______________________________ alkyl group.

7. The alkyl group represented by the formula $CH_3—CH_2—\underset{|}{C}H—CH_2—CH_3$ is classified as a __________________ alkyl group.

8. The alkyl group represented by the formula

$$CH_3—\underset{\underset{CH_3}{|}}{C}H—CH_2—$$

is classified as a __________________ alkyl group, and its isomer,

$$CH_3—\overset{|}{\underset{\underset{CH_3}{|}}{C}}—CH_3$$

is classified as a __________________ alkyl group.

9. The alkyl group represented by the formula

$$\begin{array}{c} CH_3 \\ | \\ CH_3-C-CH_2- \\ | \\ CH_3 \end{array}$$

is classified as a ________________ alkyl group.

The structural prefixes sec- (for secondary-) and tert- (for tertiary-) may be incorporated in a name to designate a specific secondary or tertiary alkyl group, respectively, if no isomeric alkyl groups of the same classification are possible. These prefixes are underlined to indicate italics and are separated from the rest of the name by hyphens. The stem used with these prefixes specifies the total number of carbons in the alkyl group.

10. The IUPAC name sec-butyl refers to the alkyl group represented by the structure.

____.

No other secondary butyl group can be drawn. Because isomeric secondary alkyl groups are possible when there are more than four carbons in unbranched alkyl groups, the prefix sec- is not definitive and is not used for any of them.

11. There is a single tertiary butyl group, which may be represented by the structure

and given the trivial name ________________.

12. For the alkyl group represented by the formula

$$\begin{array}{c} CH_3 \\ | \\ CH_3-C-CH_2-CH_3 \\ | \end{array}$$

IUPAC rules permit the trivial name, ________________, as well as the substitutive name,

________________.

If the alkyl group has a single branch at one end and the point of attachment at the other end, the structural prefix iso may be used. For example,

$$\begin{array}{l} CH_3-CH-CH_2-CH_2- \\ \quad\quad\;\; | \\ \quad\quad CH_3 \end{array}$$

may be named isopentyl. IUPAC usage of isoalkyl as a specific (trivial) name is restricted to alkyl groups with fewer than seven carbons. Similarly, the alkyl group related to the alkane neopentane may be named neopentyl. Notice that the prefixes *sec*- and *tert*- are underlined (for italics) and set off by hyphens, while the prefixes iso and neo are written without separation or underlining. This difference is an oddity of nomenclature that has little justification other than general usage. For alphabetical arrangement of substituent groups, the separated prefixes (*sec*- and *tert*-) are ignored. The initial letter of *sec*-butyl is considered to be *b*, but the initial letter of isobutyl is considered to be *i*.

13. Isobutane is the name for the alkane represented by the structural formula

________________________________,

and isobutyl is the name of the alkyl group related to it. The isobutyl group may be represented by the structural formula

____

and will be classified as a ________________ alkyl group.
(primary, etc.)

14. Isopentane may be represented by the structural formula

____

and isopentyl by the structural formula

____

The isopentyl group is classified as a ________________ alkyl group.

15. $CH_3-\underset{\displaystyle CH_3}{\underset{|}{CH}}-CH_2-CH_2-CH_2-$ is named ____________________.

16. Neopentane may be represented by the structural formula

____

and neopentyl by the structural formula

________________________.

17. The butyl groups for which the prefixes sec- and iso are appropriate are often confused by students. The butyl group for which the prefix sec- is correct contains an unbranched chain of carbon atoms; the butyl group for which the prefix iso is correct contains a ________________ chain of carbon atoms.

18. The name sec-butyl designates the structure

________________________.

19. The name isobutyl designates the structure

________________________.

20. Isobutyl is classified as a ________________ alkyl group, and sec-butyl is classified as a ________________ alkyl group.

21. The alkyl groups named sec-butyl, isobutyl, and tert-butyl are also identified by the substitutive names ________________ , ________________ , and ________________ , respectively.

22. In a departure from strictly systematic nomenclature, the group

$$\begin{array}{ccccc} CH_3 & - & CH & - & CH_3 \\ & & | & & \end{array}$$

is named isopropyl instead of <u>sec</u>-propyl. This name illustrates the concessions made to familiar, well-established names when systematization was undertaken. The group

$$\begin{array}{ccccc} CH_3 & - & CH & - & CH_3 \\ & & | & & \end{array}$$

will actually be classified as a ______________ alkyl group, but its IUPAC trivial name is ______________.

23. The compound represented by the formula

$$\square - \begin{array}{ccc} CH & - & CH_3 \\ | & & \\ CH_3 & & \end{array}$$

may be named as a substituted cyclobutane. The trivial name of the substituent is ______________, and the name of the compound is ______________________________.

24. The longest continuous chain of carbons in the formula

$$\begin{array}{ccccccccccccccccccc} CH_3 & - & CH & - & CH_2 & - & CH_2 & - & CH & - & CH_2 & - & CH_2 & - & CH_2 & - & CH_2 & - & CH_3 \\ & & | & & & & & & | & & & & & & & & & & \\ & & CH_3 & & & & CH_3 & - & CH & - & CH_2 & - & CH_3 & & & & & & \end{array}$$

contains ________ (number) carbons. Two substituents, trivially named ________ and ________, are attached to the parent chain at positions numbered ____ and ____, respectively. The ______________ (trivial name) substituent will be named first. The full name for the alkane is

is ____.

25. The alkane represented by the formula

$$\begin{array}{ccccccccccccccccc} CH_3 & - & CH_2 & - & CH_2 & - & CH_2 & - & CH_2 & - & CH & - & CH_2 & - & CH & - & CH_3 \\ & & & & & & & & & & | & & & & | & & \\ & & & & CH_3 & - & CH_2 & - & CH_2 & - & CH_2 & & & & CH_3 & & \end{array}$$

contains a parent chain of ______ carbons and a substituent whose trivial name is
(number)

__________. The substituent is located on position number ____, the name of the

parent chain is ________________, and the complete name for the alkane is __________

______________________________.

The IUPAC name of an unbranched primary alkyl group (for example, $CH_3—CH_2—CH_2—CH_2—CH_2—$, pentyl) does not use a structural prefix. (The structural prefix n-, for normal, is used by some chemists, but such usage is contrary to IUPAC rules and is roughly equivalent to the expression, "I seen.").

Alkyl groups may be attached to atoms other than carbon; for example, a hydrogen in an alkane may be replaced by chlorine. These compounds may be named in two ways: substitutive names and functional class names. The basis (parent compound) of a substitutive name can stand alone as the name of an individual compound, and modifiers (such as substituent names) must be written as part of the same word to avoid any ambiguity. For substitutive names, chlorine is regarded as a substituent (named chloro), and the one-word substitutive name for $CH_3—Cl$ is chloromethane. (Methane as a separate word could be mistaken for the name of another compound not intended.) Functional class names are often two-word names, because the final portion of the name (functional class designation) cannot stand alone as the name of an individual compound. For example, $CH_3—Cl$ is a member of the class alkyl chloride and is named with the functional class name, methyl chloride. (Chloride as a separate word will not be mistaken for the name of a specific compound.)

26. The trivial name of the alkyl group in the formula

```
CH3—CH—CH3
     |
     Cl
```

is ______________________, the functional class name for the compound is

______________________________, and the substitutive name for the compound

is ______________________________.

27. Butyl chloride may be represented by the structural formula

______________________________,

sec-butyl chloride by the formula

______________________________,

and tert-butyl chloride by the formula

________________________________.

28. Neopentyl chloride may be represented by the structural formula

____

and named with the substitutive name, ________________________________.

Neopentyl chloride is classified as a ________________ alkyl chloride.
(primary, secondary, etc.)

29. The substitutive name for

$$\begin{array}{l} CH_3-CH-CH_2-CH_2-Cl \\ \quad\quad\;\; | \\ \quad\quad CH_3 \end{array}$$

is ________________________________; the functional class name for this same compound is ________________________________. This alkyl chloride is classified as a ________________ alkyl chloride.

30. You should practice writing structural formulas from names of compounds. Consider the name 2,2,4-trimethylpentane. The parent chain name is ____________, which signifies ________ carbons in a continuous chain. Write a chain of Cs to represent this portion of the
(number)

name ________________________________.

There are ________ methyl substituents on the parent chain located at positions numbered
(number)

________, ________, and ________. Write the parent chain again, and number each of the positions:

________________________________.

Now write the parent carbon chain, and place methyl groups in the proper positions.

____.

Complete the (condensed structural) formula by writing in Hs to indicate the normal valence of 4 for each carbon.

____.

31. In the same fashion, step by step, write a formula for 3-chloro-3-isopropyl-2,4-dimethylheptane.

____.

32. There are three isomeric secondary alkyl chlorides with molecular formula $C_5H_{11}Cl$, which may be represented by the structural formulas

____________________, ____________________, and ____________________.

33. There are three isomeric tertiary alkyl chlorides with molecular formula $C_6H_{13}Cl$, which may be represented by the structural formulas

____________________, ____________________, and ____________________.

Because of the existence of isomers, as illustrated in items 32 and 33, sec-pentyl chloride and tert-hexyl chloride are not definitive names for any compounds and are therefore unacceptable. Each of the six alkyl chlorides in items 32 and 33 may, however, be named with a substitutive name.

34. A substitutive name for

$$\begin{array}{l} CH_3-\underset{\displaystyle Cl}{\underset{|}{CH}}-\underset{\displaystyle CH_3}{\underset{|}{CH}}-CH_3 \end{array}$$

is ______________________.

35. 2-Chloro-2,3-dimethylbutane may be represented by the structural formula

______________________.

3

Nomenclature of Alkenes

SUBSTITUTIVE NAMES

Alkenes contain a double bond between a pair of carbon atoms, and the systematic ending to be used in names of alkenes is ene. Ethene is a compound containing two carbons (stem eth) and a carbon-carbon double bond (ending ene). The rules for naming alkanes apply to alkenes with two additional restrictions: The chain chosen as the basis for the name is the longest continuous chain containing the carbon-carbon double bond (the functional group), and the parent chain is numbered to assign the functional group the smaller possible number. The parent chain may or may not be the longest continuous chain of carbon atoms in the compound. The position of the functional group is designated by the lower number assigned to the carbons joined by the double bond; this number generally immediately precedes the stem of the name. For example, $CH_2{=}CH{-}CH_2{-}CH_3$ is named 1-butene.

1. The parent chain in the compound

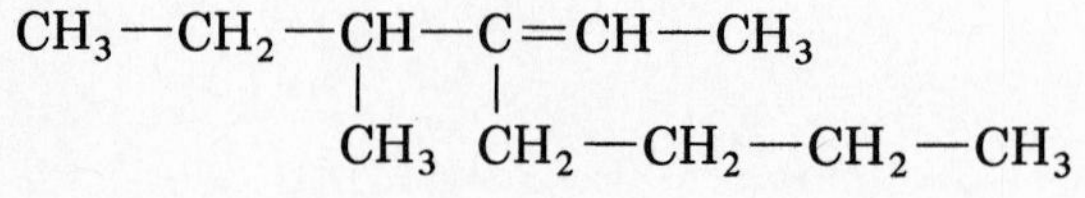

$$
\begin{array}{llllll}
CH_3{-}CH_2{-} & CH{-} & C{=}CH{-}CH_3 & & & \\
 & | & | & & & \\
 & CH_3 & CH_2{-}CH_2{-}CH_2{-}CH_3 & & &
\end{array}
$$

contains ______ carbon atoms.

2. The alkene name for this parent chain is ______________.

3. The substituent on the parent chain is named ______________________.

4. When the parent chain is properly numbered, the functional group is assigned the position number ______, and the substituent is assigned the position number ______.

5. The complete designation of the parent chain, including locant for the functional group, is ______________.

6. The complete name for the formula is ______________________________.

7. The name for the parent chain in the formula

$$CH_3-\underset{\underset{CH_3}{|}}{\overset{\overset{CH_3}{|}}{C}}-CH_2-\overset{\overset{CH_3}{|}}{CH}-CH=CH_2$$

is ______________.

8. The functional group is assigned locant ____, and the substituents are assigned locants ____, ____, and ____.

9. Locants 2, 2, and 4 for the substituents are incorrect because ______________________________
____
____
____.

10. A complete name for the formula is ______________________________.

11. On the basis of only differences in carbon skeleton and in location of the C=C, the formulas for the isomeric alkenes containing four carbons are

____________________, ____________________, and ____________________.

12. These alkenes are named, respectively, ______________, ______________, and ____________________.

Cyclic alkenes are called <u>cycloalkenes</u>, and the alkene linkage is always assigned the locant 1. The locant is unnecessary in the name.

13. A name for

[octagon ring structure with one double bond]

is ____________________.

14. A structural formula for cyclopentene is

________________________.

15. A structural formula for 4-_tert_-butylcyclohexene is

______________________________.

16. The compound named 1-_sec_-butylcyclopentene may be represented by the formula

____________________________________.

17. Halogens and other substituents are treated exactly as they are in names for substituted alkanes. The substituent in the formula $CH_2{=}CH{-}CH_2{-}Cl$ is located on position

________ of propene, and the substitutive name for the compound is

____________________________________.

18. The formula

$$\begin{array}{l} \quad -Cl \qquad CH_3 \\ \qquad\qquad\quad\;\; | \\ \quad -CH_2-CH-CH_3 \end{array}$$

shows substituents on positions numbered ______ and ______, regardless of the direction of numbering. The choice is made to give the smaller locant (position number) to the substituent named first in the name of the compound. The compound is named

____.

When alternative directions of numbering produce _different_ sets of numbers to designate locations of substituents, the alternative sets are compared term by term. The set with the lower number at the point of first difference in the two sets is the correct one.

19. Numbering of the ring in the formula

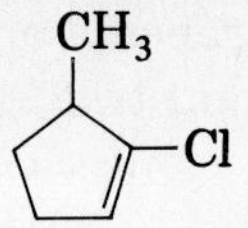

must assign numbers 1 and 2 to the carbons in the alkene linkage. If the carbon bound to chlorine is assigned position number 1, the two substituents will be located on positions numbered _____ and _____. If the other alkene carbon is assigned position number 1, the substituents will be on positions numbered _____ and _____. The two sets differ in the first locant for a substituent, and the set with the smaller first locant designates positions ____ and ____. The correct substitutive name for the substituted cycloalkene is ________________________________.

20. Substituent positions in the formula

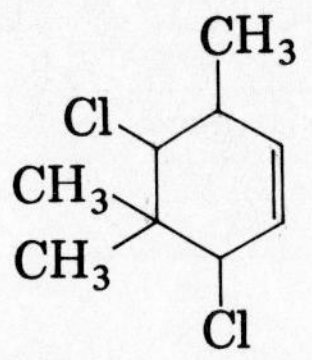

may be designated by alternative sets of locants, which are ______________ and ______________. The two sets differ first in the third locant, and the set with the smaller third locant is ______________. The correct name for the compound represented by the formula is ______________________________.

GROUPS RELATED TO ALKENES

An alkyl group can be considered to be formed from an alkane by loss of a hydrogen. In the same way, alkenyl groups are related to alkenes. In alkenyl groups, as in alkyl groups, the position from which the hydrogen can be considered to have been lost is always position number 1, and that number does not need to be included in the name. The stem appropriate to the parent chain, which includes the carbon-carbon double bond, is combined with the ending enyl; a locant for the alkene linkage immediately precedes the stem in the name. For example, $CH_2{=}CH{-}CH_2{-}$ is named 2-propenyl; the en is at position 2, and the yl is at position 1.

21. The name for the group $CH_3{-}CH{=}CH{-}$ is ______________________________.

22. Ethenyl is the name for the group ______________________.

Alternative, IUPAC-approved trivial names for ethenyl and 2-propenyl are vinyl and allyl, respectively. Vinyl chloride and allyl chloride are important industrial chemicals, and the trivial names are widely used.

23. The formula for vinyl chloride is ______________________.

24. The formula for allyl chloride is ______________________.

25. The allyl group will be classified as a ____________ group.
(primary, etc.)

26. The compound represented by the formula

[cyclohexene ring]$-CH=CH_2$

may be named as a substituted cycloalkene. The substituent, ____________, is at position
(trivial name)

__________, and the name of the compound is ______________________.

27. The functional class, trivial name for $CH_2=CH-CH_2-Br$ is ______________________

______________________.

Vinyl and allyl are not correctly used as the parent names for more complex alkenyl groups, but ethenyl and 2-propenyl are.

28. The name for the alkenyl group

$$\begin{array}{l} CH_2=C-CH_2- \\ \quad\quad\;\; | \\ \quad\quad\; CH_3 \end{array}$$

is ______________________.

29. The isomeric group

$$\begin{array}{r} CH_3 \\ | \;\;\; \\ CH_3-CH=C- \end{array}$$

is named ______________________.

30. The group 1-methyl-3-cyclopentenyl may be represented by the formula

____

and is classified as a __________ (primary, etc.) group.

CIS,TRANS ISOMERS

31. On the basis of only differences in carbon skeleton and in location of the C=C, the formulas for the isomeric alkenes containing four carbon atoms are

____________________, ____________________, and ____________________.

Because of the restriction of rotation about a carbon-carbon double bond, some alkenes can exist as spatial isomers. The distinction between these isomers depends on the spatial arrangement of the groups attached to the C=C rather than on the position of substituents or functional groups along the parent chain. Such isomerism is one kind of stereoisomerism; the designation, cis,trans isomerism, is often used. For example,

Cl   Cl
  C=C
H   H

and

Cl   H
  C=C
H   Cl

are cis,trans isomers. The two atoms or groups attached to each carbon of the alkene linkage must be different for cis,trans isomerism to be possible. That is, if one of the alkene carbons is attached to two identical atoms or groups, cis,trans isomerism will not be possible for that compound.

32. Of the alkenes represented by the formulas in item 31, only __________ (number) can exhibit cis,trans isomerism, namely __________.

33. Formulas that reveal the cis,trans isomerism of 2-butene are

____________________ and ____________________.

The adjectives *cis* and *trans* are used to differentiate cis,trans isomers. Cis designates the isomer with reference groups or atoms on the same side (side, not end) of the alkene linkage; trans designates the isomer with reference groups or atoms on opposite sides of the alkene linkage. Cis and trans designations for alkenes refer to the extension of the parent chain from the alkene linkage.

34. The cis isomer of 2-butene may be represented by the formula

______________________________.

35. The trans isomer of 2-butene may be represented by the formula

______________________________.

The prefixes *cis-* and *trans-* are included in completely descriptive names of alkenes. These italicized prefixes are separated from the rest of the name by hyphens, and they immediately precede the number indicating position of the alkene linkage. (When cis and trans are used as adjectives and not as prefixes in names, they are not italicized in general use.)

36. The structural formula for *trans*-2-pentene is

______________________________.

37. The prefix *trans-* specifies that ______________________________

______________________________.

38. A completely descriptive name for

```
                        H
                       /
CH3—CH—CH2—C          H
    |        \\      /
    CH3        C
               |
               CH3
```

is ______________________________.

39. A structural formula for 3,4-dichloro-9-methyl-trans-3-decene is

____.

Cycloalkenes with ring sizes smaller than eight members have been isolated only as cis-cycloalkenes. Cis,trans isomers of cycloalkenes with eight-membered rings and larger have been isolated. For convenience, cycloalkenes are usually represented by geometric figures, as are cycloalkanes. A cis alkene linkage is represented by \=/ (extensions of parent chain from the same side of C=C), and a trans alkene linkage by \=\ (extensions of parent chain from opposite sides).

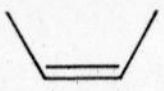

40. In a line drawing (geometric figure) formula for each of the two cyclooctenes, the fragment of the formula that represents the cis alkene linkage will look like ______________, and the fragment that represents the trans alkene linkage will look like ______________. cis-Cyclooctene may be represented by the formula

______________________,

and trans-cyclooctene by the formula

______________________.

41. The geometry indicated for the alkene linkage by the formula

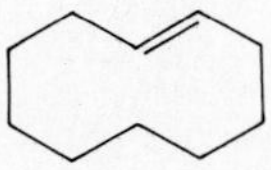

is ______________ (cis or trans). The ring contains ______________ (number) carbons, and a complete name for the compound illustrated is ______________________________.

The configuration (spatial arrangement) of some isomers, such as

$$\begin{array}{ccc} CH_3 & & Cl \\ & C{=}C & \\ CH_3{-}CH_2 & & Br \end{array} \quad \text{and} \quad \begin{array}{ccc} CH_3 & & Br \\ & C{=}C & \\ CH_3{-}CH_2 & & Cl \end{array},$$

is not readily specified by the labels cis and trans. A more general specification of configuration of alkenes has been devised and is widely used, even in place of cis and trans with simple alkenes. This specification depends on assigning higher or lower priority to each of the two atoms or groups bound to each carbon in the alkene linkage. If the two higher-priority atoms or groups are on the same side of the C=C, the label (prefix) is *Z* (for German *zusammen* = together); if they are on opposite sides, the label (prefix) is *E* (for German *entgegen* = opposite).

Priorities of the atoms or groups in each pair are assigned by application of a comprehensive set of rules. Three rules selected from that set are likely to cover most alkenes in a beginning course.

(1) The higher the atomic number of the atom directly attached to the alkene carbon, the higher the priority of that atom (or group). For example, CH_3- is higher priority than H— (atomic number 6 vs. 1), but Cl— is higher than CH_3- (17 vs. 6).

(2) For isotopes of the same element, the higher priority goes to the one of higher mass number (D higher than H).

(3) If the two atoms attached directly to the alkene carbon are the same, the one attached in turn to atoms of higher atomic number has the higher priority. For example, CH_3-CH_2- is higher priority than CH_3- (C attached C,H,H vs. C attached to H,H,H), but $(CH_3)_2CH-$ is higher priority than CH_3-CH_2- (C attached to C,C,H vs. C attached to C,H,H).

In names, the prefix *E* or *Z* appears first, is underlined (italicized), enclosed in parentheses, and separated by a hyphen from the rest of the name: for example, (*E*)-2-pentene.

42. In the formula

$$\begin{array}{ccc} Cl & & CH_3 \\ & C{=}C & \\ H & & CH_2{-}CH_3 \end{array}$$

the two atoms attached to the left carbon in the alkene linkage (C=C) have the priorities: higher ______, lower ______. This priority order is assigned on the basis of __________

____.

The two atoms attached to the right carbon in the C=C are the same, but priorities of the groups can be assigned on the basis of ____

____.

For those groups, the priorities are: higher ______________, lower ______________.

The two higher-priority atoms/groups appear on the ______________________________

side(s) of the C=C, and the configuration of this alkene is ________ (E or Z). The parent alkene of

this compound is named ______________________________, and the full name of the

compound illustrated, including specification of configuration, is ______________________

____.

43. The configuration of

$$\begin{array}{ccc} CH_3{-}CH_2 & & CH_2{-}CH_3 \\ & \diagdown\;\;\;\diagup & \\ & C{=}C & \\ & \diagup\;\;\;\diagdown & \\ CH_3{-}CH_2{-}\underset{\displaystyle CH_3}{\underset{|}{CH}} & & CH_3 \end{array}$$

may be specified as E or Z by assigning priorities to the two pairs of alkyl groups attached to the alkene carbons. Besides an alkene carbon, the first carbon in each alkyl group is attached in turn to three other atoms. In ethyl, they are ______, ______, and ______; in sec-butyl, ______, ______, and ______; and in methyl, ______, ______, and ______. Ethyl has ______________ (higher or lower) priority than does sec-butyl and ______________ (higher or lower) priority than does methyl. The configuration of this alkene is therefore ____________, and the full name for the alkene is ____.
If the alkene were designated cis or trans, the correct designation would be ____________.

44. The configuration of the alkene represented by the formula

$$\begin{array}{ccc} CH_3{-}CH_2{-}CH_2 & & CH_3 \\ & C{=}C & \\ (CH_3)_3C & & H \end{array}$$

is ______ (E or Z) or ____________ (cis or trans). (Note that E, Z and cis,trans here and in item 43 come from the use of different reference groups, that E and trans are not always associated with the same structure, and that the prefixes would appear at different points in names of the alkenes.)

45. (Z)-2—Chloro-1-cyclopropyl-1-butene may be represented by the formula

____.

46. The compound

$$\begin{array}{ccc} CH_3 & & H \\ & C{=}C & \\ D & & D \end{array}$$

has the __________ configuration.
(E or Z)

DIENES

Hydrocarbons containing two separate carbon-carbon double bonds are named as alkadienes. The appropriate stem signifying the number of carbon atoms in the parent chain is combined with the ending adiene. To learn why the "a" is included in the ending, try saying aloud "pentdiene" and "pentadiene." The "a" simply makes the word easier to pronounce. The parent chain is numbered to give the lower possible locants to the double bonds; a locant for each double bond precedes the stem of the name.

47. 1,4-Pentadiene may be represented by the structural formula ________________________

____________________, and its isomer, 1,3-pentadiene, by the formula ________________________

____________________.

48. The hydrocarbon represented by the formula $CH_2{=}CH{-}CH{=}CH_2$ is probably the most important diene from the standpoint of industrial use; it is named ________________________

____________________.

49. Isoprene is the common (trivial) name accepted by IUPAC for 2-methyl-1,3-butadiene, which may be represented by the formula

____.

Isoprene is regarded as the building block for many complex compounds occurring in nature.

50. Propadiene, ________________, is also known as allene. Allene is accepted as an
(structural formula)
IUPAC name for this unsubstituted diene, but the more systematic name, propadiene, is preferred.

51. The name of the compound $CH_3—CH=C=CH_2$ is ________________.

52. 1,3-Cyclohexadiene may be represented by the formula

________________.

1,2-Dienes are classified as cumulated dienes, 1,3-dienes as conjugated dienes, and dienes with greater separation of the two double bonds as isolated dienes.

53. Isoprene (2-methyl-1,3-butadiene) is classified as a(n) ________________ diene.

54. The hydrocarbon represented by the formula

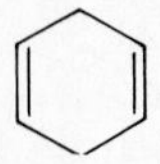

is classified as a(n) ________________ diene.

55. Propadiene, represented by the structural formula ________________, is classified as a(n) ________________ diene.

56. 2-Chloro-5-ethyl-3,5-decadiene, represented by the structural formula

________________________________,

is classified as a(n) ________________ diene. A formula that illustrates the (3Z, 5Z) configuration for this diene is

________________________________.

(Note: For priority assessment, each bond from C is considered separately; the group fragment C=C is considered as C attached to C,C.)

Hydrocarbons containing more than two separate carbon-carbon double bonds are named in similar fashion, di being replaced by the appropriate multiplying prefix.

57. 1,3,5,7-Cyclooctatetraene has ________ (number) C=C and can be represented by the line-drawing formula

________________.

BIVALENT GROUPS

In the same way that one can imagine the formation of a univalent group (alkyl) by removal of a hydrogen from an alkane, one can imagine the formation of a bivalent group by removal of two hydrogens from an alkane. As with the names of alkyl and alkenyl groups, the names of bivalent groups may be used in functional class names of compounds. Except for rather simple structures, however, such names are seldom encountered. The simplest bivalent group, $—CH_2—$, is named methylene, and CH_2Cl_2 is often named methylene dichloride.

58. The bivalent group, $—CH_2—CH_2—$, may be called dimethylene or ethylene. Ethylene dibromide is an acceptable IUPAC name for $Br—CH_2—CH_2—Br$, but that compound is indexed in *Chemical Abstracts* only by its substitutive name, ________________________.

The bivalent names offer no advantage over the systematic, substitutive names, and their use seems to be decreasing.

4
Alkynes

Alkynes contain a triple bond between a pair of carbon atoms somewhere in the molecule and are named in much the same way as are alkenes. The systematic ending yne, signifying the triple bond between a pair of carbon atoms in the parent chain, is combined with the appropriate stem, and the position of the C≡C is indicated by the smaller possible locant, which generally precedes the stem of the name.

1. The name for HC≡CH is ethyne, and the name for $CH_3—C≡CH$ is ______________. Ethyne is also known as acetylene, an alternative name accepted by IUPAC. There is little other than familiarity to recommend usage of this older name, however, and *Chemical Abstracts* does not use it for indexing. Ethyne is shorter and systematic. Substituted ethynes should be named as ethynes, not as acetylenes.

2. In the formula

$$\begin{array}{l} CH_3—\underset{\displaystyle CH_3}{\underset{|}{CH}}—C{\equiv}C—\underset{\displaystyle CH_2—CH_3}{\underset{|}{CH}}—CH_2—CH_3 \end{array}$$

the longest continuous chain containing the functional group, C≡C, contains ____________ (number) carbons; the stem signifying this number of carbons is ________. The ending signifying the functional group, C≡C, is ________, and the name for the chain containing this group is ____________. When the chain is properly numbered, the functional group will be assigned position number ____, and the parent name, including locant, will be ____________. The two substituents, ________ and ________, are on positions ____ and ____, respectively, and the full name for the compound represented by the formula is ______________________________.

3. The name for $CH_3—C≡C—CH_3$ is ______________________.

4. The structural formula for 3-hexyne is ______________________________________.

5. The structural formula for 4-sec-butyl-1-chloro-2-octyne will contain a parent chain of

______ carbons, signified by the stem ______. The ending, yne, indicates the functional
(number)

group ______. The parent chain can then be written ______________. The substituent on position 1 is represented by the symbol ______, the one on position 4 by

______________________.

The complete structural formula for 4-sec-butyl-1-chloro-2-octyne may be drawn as

______________________________. The systematic, substitutive name for the substituent on position 4 is __________, and the name for the compound illustrated, using this substituent name, is

______________________________________.

MULTIPLE MULTIPLE BONDS

Compounds containing more than one C≡C may be named as alkadiynes, alkatriynes, and so forth. The names parallel those you have studied for alkadienes.

6. 3-Isobutyl-1,4-hexadiyne may be represented by the formula

______________________________. When the systematic, substitutive name is used for the substituent, this compound will be named

______________________________.

7. The parent compound on which the name of

```
           CH2—C≡C—CH—CH3
          /              \
CH3—CH                    CH—CH3
          \              /
           CH2—C≡C—CH2
```

will be based is ________________________________. Correct numbering of the parent ring will assign to the substituted positions the numbers ______, ______, and ______. The complete name of the compound illustrated is

____.

IUPAC substitutive names for compounds containing both a C=C and a C≡C are based on the general parent name "alkenyne." The parent chain is numbered so as to use the lower locants for the double and triple bonds. For example, $CH_3{-}CH{=}CH{-}C{\equiv}CH$ is named 3-penten-1-yne, <u>not</u> 2-penten-4-yne. Note that the locant for C=C (en) precedes the stem and that for C≡C (yne) immediately precedes the yne ending. This style will be used whenever two systematic endings are used in the same name. If two directions of numbering the parent chain of an alkenyne use the same locants for the two functional groups, the lower number for the double bond is used, even though yne is the final ending of the name. For example, $CH_3{-}CH{=}CH{-}CH_2{-}C{\equiv}C{-}CH_3$ is named 2-hepten-5-yne, <u>not</u> 5-hepten-2-yne.

8. The compound $CH_2{=}CH{-}C{\equiv}CH$ is an important industrial chemical which is commonly called <u>vinylacetylene.</u> IUPAC does not approve the use of "acetylene" as the parent on which a substitutive name is built, and the IUPAC-approved name for this compound is

______________________.

9. The name of the compound represented by the formula

$$\begin{array}{ccccccccc} & & Cl & & & & CH_3{-}CH{-}CH_3 & & \\ & & | & & & & | & & \\ CH_3{-} & C{\equiv}C{-} & CH & {-}CH_2{-} & CH{=} & & C & {-}CH_2{-} & CH_3 \end{array}$$

is ____,
and the name of its isomer,

$$\begin{array}{ccccccccc} & & Cl & & & & CH_3{-}CH{-}CH_3 & & \\ & & | & & & & | & & \\ CH_3{-} & C{\equiv}C{-} & CH & {-}CH_2{-} & CH_2{-} & & C & {=}CH{-} & CH_3 \end{array}$$

is ____.

10. The name of

$$\begin{array}{ccc} & CH_3 & \\ & | & \\ CH_2{=}CH{-}CH_2{-} & C & {-}C{\equiv}CH \\ & | & \\ & CH_2{-}CH_3 & \end{array}$$

is ________________________________.

ALKYNYL GROUPS

The formulas and names of alkynyl groups are related to those of alkynes in the same way as formulas and names of alkenyl groups are related to those of alkenes.

11. The group HC≡C— is often regarded as a substituent and is named

________________________.

12. 1-tert-Butyl-4-ethynylcyclohexene may be represented by the formula

________________________.

13. The formula for the 2-propynyl group is ________________, and that for the 1-propynyl group is ________________. Remember: The point of attachment of the alkyl, alkenyl, or alkynyl group is always the number 1 position of that group.

14. Three isomeric groups can be formed by the removal of one hydrogen from 1-penten-3-yne. The formulas for these three groups are ________________, ________________, and ________________, and their names are ________________________, ________________________, and ________________________, respectively.

5
Aromatic Hydrocarbons

Benzene, the parent aromatic hydrocarbon, is symbolized in various ways:

C_6H_6,

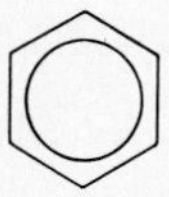

The orientation of the hexagon does not matter.

Benzene serves as the basis of the substitutive names of many substituted benzenes. R is often used as a general symbol for an alkyl group. The one-word name for an alkyl-substituted benzene such as

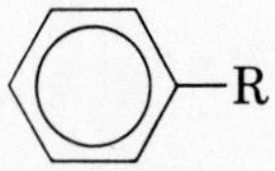

—R or $C_6H_5—R$

is formed by following the name of the alkyl group with benzene (for example, methylbenzene). Since all positions in benzene are equivalent, no number is needed in the name to indicate the position of a single substituent.

1. The aromatic hydrocarbon represented by the formula $C_6H_5—CH_2—CH_3$ or

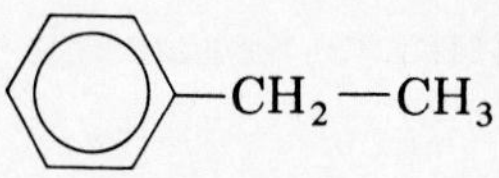

—$CH_2—CH_3$

will be named______________________________.

2. The ethynyl group may be represented by the formula____________________, and ethynylbenzene may be represented by the formula

____________________________________.

3. sec-Butylbenzene may be represented by the formula

______________________________.

4. The alkyl group attached to C_6H_5- in the formula

$$C_6H_5-CH_2-\overset{\displaystyle CH_3}{\underset{\displaystyle CH_3}{\overset{|}{\underset{|}{C}}}}-CH_3$$

has the trivial name______________________________, and the substituted benzene is named______________________________________.

The class of compounds of which benzene and substituted benzenes are members is often called arene. This general term is related to benzene as alkane is related to methane.

Sometimes a chemist may prefer to regard the benzene portion of the molecule as a substituent rather than as the parent compound. Such an occasion may occur when a list of related compounds is being compiled. As a substituent, C_6H_5- or

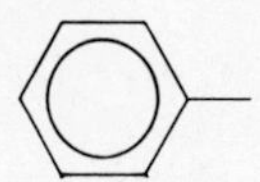

is named phenyl and is treated as any other substituent. (The completely systematic name is benzenyl, but this name has never challenged phenyl in usage.) Phenyl is an example of an aryl substituent, as methyl is an example of an alkyl substituent. For convenience, phenyl is frequently symbolized ϕ or Ph.

5. $(C_6H_5)_3CH$ is most easily named as a substituted methane; its name then is

______________________________________.

6. Ethynylbenzene, represented by the formula

____________________________,

may also be named as a substituted alkyne. The name on that basis will be

__________________________________. Names based on the larger parent compound are generally preferred.

7. An IUPAC name for

$$\begin{array}{l} CH_3{-}CH{=}CH{-}\underset{\displaystyle CH_3}{\underset{|}{CH}}{-}CH_2{-}\underset{\displaystyle CH_2{-}CH_2{-}CH_3}{\underset{|}{CH}}{-}\phi \end{array}$$

with phenyl being treated as a substituent like methyl, will use as the basis of the name a hydrocarbon chain named ______________. The methyl substituent appears on position number _____, and the phenyl substituent on position number _____. The complete IUPAC name for the substituted alkene is ____.

8. A formula for 5-phenyldodecane may be drawn as

____.

9. Allylbenzene, represented by the formula ______________________________, may be named as a substituted alkene: ____. Its isomer, (1-propenyl)benzene, represented by the formula ______________________________, may also be named as a substituted alkene: ____.

When two or more substituents are attached to the benzene ring, isomers are possible, and position designations must be used. Two different position designations are used: numbers and letters.

When numbers are used, one of the substituted positions will always be numbered 1, and the other positions in the ring are numbered 2 through 6. The number 1 position and the direction of numbering are chosen so that the smallest set of numbers is assigned to the substituted positions.

10. In the formula

CH_3–(benzene ring)–CH_3

the methyl substituents appear on positions numbered _____ and _____.

11. 1,2-Diethylbenzene may be represented by the formula

______________________________.

12. The formula

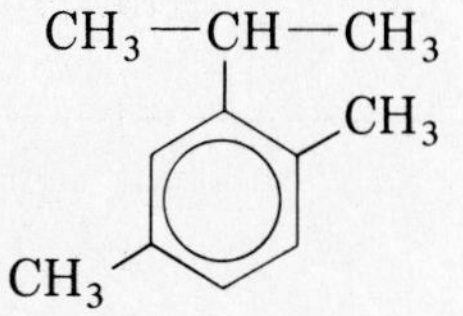

$CH_3-CH-CH_3$

CH_3

CH_3

pictures substituents on positions numbered ______, ______, and ______ (the lowest numbers that can be used). The name of the compound represented by this formula is

____.

13. Mesitylene is a common name sometimes used for 1,3,5-trimethylbenzene,

______________________________.
(formula)

For most disubstituted benzenes, chemists frequently use letters rather than numbers to designate positions of substitution: o- (standing for and read ortho-) is used for compounds in which substituents appear on positions numbered 1 and 2; m- (for meta-) signifies substituents on positions 1 and 3; and p- (for para-) signifies substitutents on positions numbered 1 and 4. These letter designations are prefixes which are italicized and set off from the rest of the names by hyphens.

14. Xylene is a common name for dimethylbenzene. There are three isomeric xylenes known:

o-xylene, ______________________________,
(formula)

m-xylene, ____________________ , and
(formula)

p-xylene, ____________________ .
(formula)

15. 1,4-Diisopropylbenzene may also be called ____

16. The two alkyl substituents in the formula

$$C_6H_5\text{ (ring) with } -CH_2-CH_3 \text{ and } -CH(CH_3)-CH(CH_3)-CH_3 \text{ on adjacent carbons}$$

are ____________ to each other.
(o-, m-, or p-)

17. m-Ethylvinylbenzene can be represented by the formula

____________________________________ .

18. 1-Hexyl-2-isobutylbenzene may also be named ____________________________________ .

Methylbenzene is usually called toluene (even though *Chemical Abstracts* indexes the compound as methylbenzene), and toluene is actually used as the basis of IUPAC names. That is, toluene is treated as a parent compound just as benzene is. When toluene is used in this way, the carbon to which the methyl group is attached is position number 1.

19. 3-Ethyltoluene may be represented by the formula

____________________________________ .

20. p-tert-Butyltoluene may be represented by the formula

____.

21. 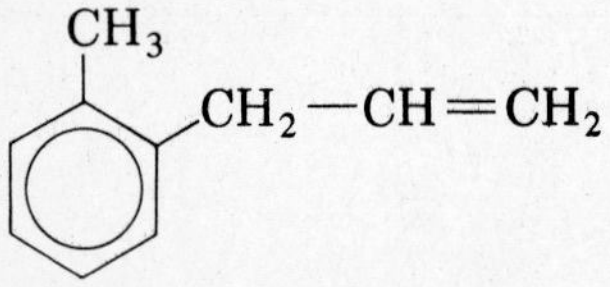 [structure: benzene ring with CH_3 and, ortho to it, $CH_2-CH=CH_2$] may be named as a substituted toluene, ____________________.

22. Named as a substituted benzene,

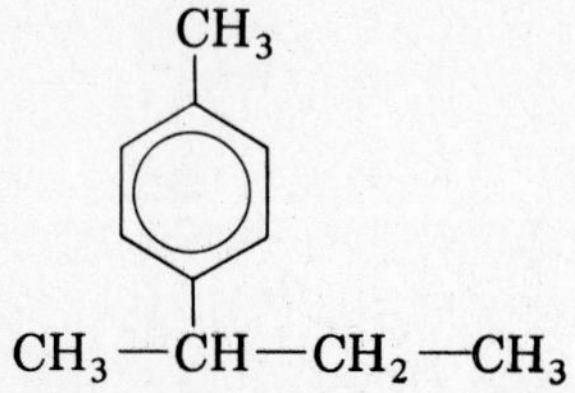

[structure: benzene ring with CH_3 at top and, para to it, $CH_3-CH-CH_2-CH_3$ attached through CH]

will be called ____; named as a substituted toluene, the same compound will be called ____.

Some other substituted benzenes are also treated as parent compounds like toluene.

23. For example, ethenylbenzene, represented by the formula

____,

is most often called styrene. *Chemical Abstracts*, however, uses only ethenylbenzene for indexing purposes.

24. p-Isobutylstyrene may be represented by the formula

____.

The two positions in the vinyl group of styrene have been designated by the Greek letters alpha and beta, as illustrated in the following formula:

$$C_6H_5-\underset{\alpha}{CH}=\underset{\beta}{CH_2}$$

This style of position designation, although permitted, is no longer recommended. Names based on benzene as the parent compound are preferred and are the ones used by *Chemical Abstracts.*

25. The compound represented by the formula

$$C_6H_5-C(CH_3)=CH_2$$

has been called α-methylstyrene, but it should be named as a substituted benzene,

______________________________;

An isomer, (1-propenyl)benzene, may be represented by the formula

______________________________;

Another isomer, 3-methylstyrene, may be represented by the formula

______________________________.

26. The substituted styrene

$$C_6H_5-C(=CH_2)-CH_2-CH_2-CH_3$$

may be named as a substituted alkene with the name ______________________________; or it may be named as a substituted benzene (*Chemical Abstracts*). The choice of name often depends upon the context in which one wishes to use the name.

27. Phenylbenzene, represented by the formula

______________________________,
is usually called biphenyl. Biphenyl is an IUPAC-approved parent compound name. The positions in each ring are numbered 1 through 6, beginning at the carbon bonded to the other ring; one set of locants is primed (1′ through 6′) to distinguish that ring from the other. For example,

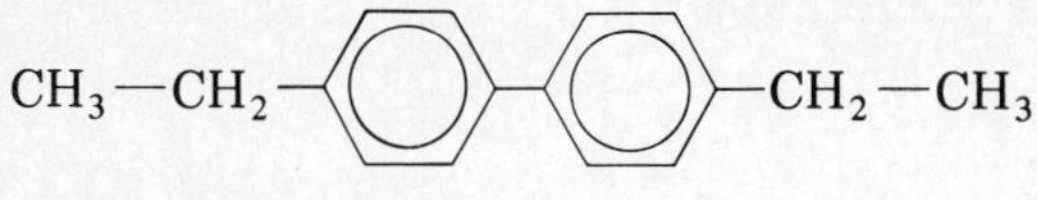

CH_3-CH_2- $-CH_2-CH_3$

is named 4,4′-diethylbiphenyl.

28. The compound

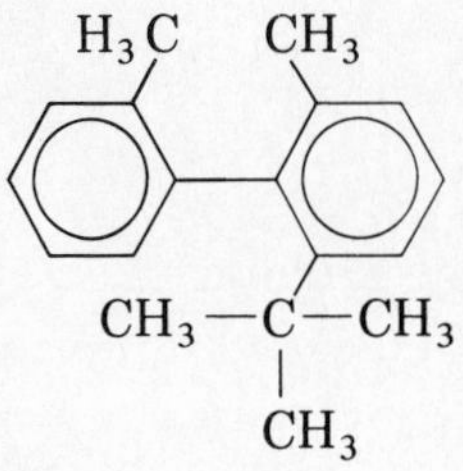

H_3C CH_3

CH_3-C-CH_3

CH_3

is named ____.

A polycyclic arene which is the parent compound of a number of industrial products is naphthalene, $C_{10}H_8$,

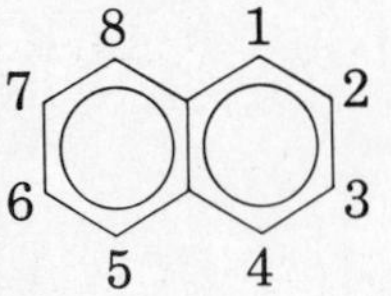

8 1
7 2
6 3
5 4

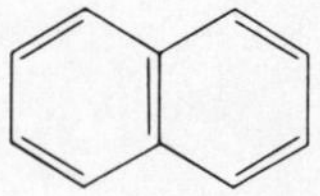

The locants for the positions shown in the formula. The aryl groups related to naphthalene,

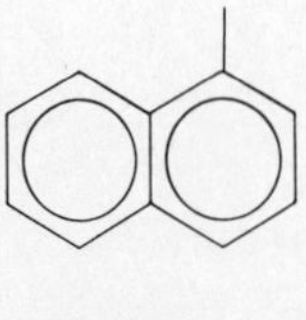

and

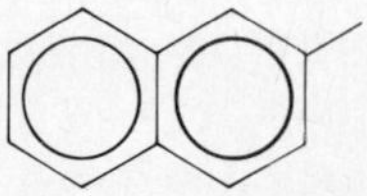

are named 1-naphthyl and 2-naphthyl (*Chemical Abstracts* uses the longer names 1- and 2-naphthalenyl). Note that established, invariant numbering of polycyclic compounds such as naphthalene permits and requires a locant other than 1 for some of the related groups.

29. 1-Methylnaphthalene may be represented by the formula

______________________________,

2-phenylnaphthalene by the formula

______________________________,

and 1,7-dimethylnaphthalene by the formula

______________________________.

30. The compound

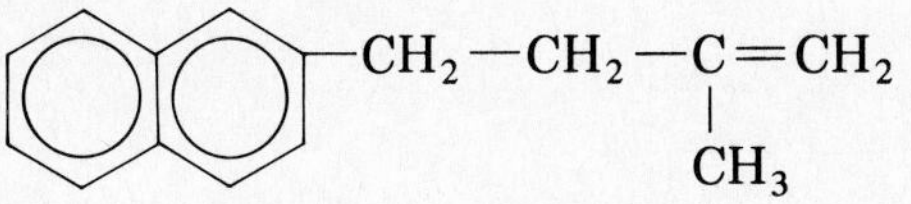

$$C_{10}H_7{-}CH_2{-}CH_2{-}\underset{\displaystyle CH_3}{\underset{|}{C}}{=}CH_2$$

may be named as an alkene or as a naphthalene. As an alkene, it will be named

____, and as a naphthalene, it will be named ____.

Hydrogens added to an unsaturated parent compound (such as an arene) may be indicated in a substitutive name (based on the parent compound) by the prefix hydro together with the appropriate multiplying prefix and locants. (Hydro means "hydrogen added" rather than "hydrogen substituted.") For example,

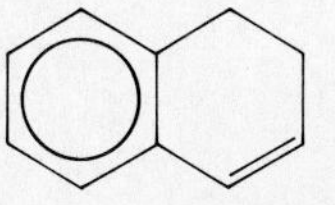

or

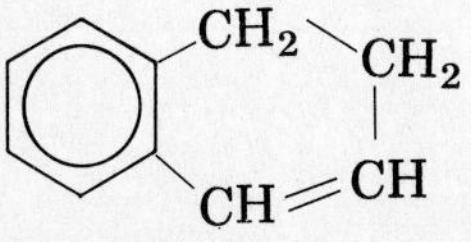

CH_2 CH_2 $CH{=}CH$

is related to naphthalene by the addition of hydrogens at positions 1 and 2 and is usually named 1,2-dihydronaphthalene. Note that the components of the name ($2H + C_{10}H_8$) add up to the correct composition of the compound ($C_{10}H_{10}$).)

31. 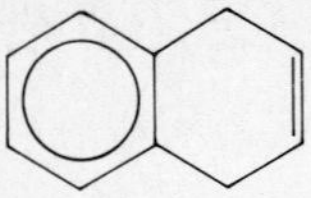 , related to naphthalene by addition of hydrogens at positions ____ and ____, may be named ______________________________; , related to naphthalene by addition of hydrogens at positions __________, may be named ______________________________; and 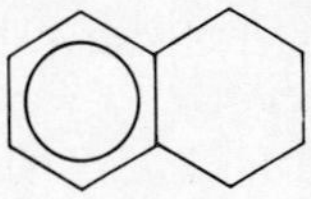 may be named ______________________________.

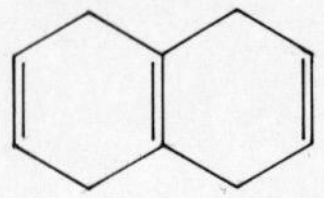

6
Alcohols

Alcohols are compounds in which an alkyl group is attached to an —OH group and may be represented by the general formula R—O—H (R is a convenient symbol for an unspecified hydrocarbon group). IUPAC names for alcohols are usually one-word substitutive names or two-word functional class names. Both kinds of names are used by chemists, especially for some of the low-molecular-weight alcohols for which the functional class name may be the more familiar one. *Chemical Abstracts*, with its emphasis on indexing and a single name for a compound, however, does not use the functional class names for indexing.

FUNCTIONAL CLASS NAMES

The functional class of organic compounds with the general formula R—OH is alcohol. Alcohol is a classification term, not the name of any individual compound, and it is treated in much the same fashion as the classification term chloride. The name of the alkyl group attached to —OH is specified, and the name of the compound is completed by the separate word alcohol. For example, CH_3—OH is methyl alcohol.

1. The functional class name for

$$\begin{array}{c} CH_3-CH-CH_3 \\ | \\ OH \end{array}$$

is ______________________________.

2. The structural formula for sec-butyl alcohol is

______________________________.

3. Functional class names of alcohols are two-word names because ______________

______________________________.

4. The trivial name for the group

$$\begin{array}{l} CH_3-CH-CH_2-CH_2- \\ \quad\quad\;\; | \\ \quad\quad\; CH_3 \end{array}$$

is ____________ , and the functional class name for the compound represented by the formula

$$\begin{array}{l} CH_3-CH-CH_2-CH_2-OH \\ \quad\quad\;\; | \\ \quad\quad\; CH_3 \end{array}$$

is ______________________________.

5. The IUPAC trivial name for the group $CH_2=CH-CH_2-$ is ____________, and the functional class name for $CH_2=CH-CH_2-OH$ is ______________________________.

6. The condensed structural formula for tert-butyl alcohol is

______________________________.

On the same basis as is used for alkyl chlorides, alcohols can be classified as primary, secondary, and tertiary. For example, if the carbon attached to the OH group is attached to only one carbon, the alcohol is a primary alcohol.

7. Ethyl alcohol, represented by the structural formula ____________________, is classified as a ______________ alcohol.

8. Isobutyl alcohol, represented by the structural formula

______________________________,

is classified as a ______________ alcohol.

9. Allyl alcohol is a ______________ alcohol.

10. These three alcohols are all primary alcohols because ______________________________

____

____.

11. Isopropyl alcohol is classified as a __________________ alcohol.

12. A secondary alcohol whose molecular formula is C_4H_9OH has the structural formula

____

and the functional class name ______________________.

13. There are three secondary alcohols with molecular formula $C_5H_{11}OH$; their structural formulas are

____________________, ____________________, and ____________________.

14. The tertiary alcohol of lowest molecular weight has the structural formula

____

and the functional class name ______________________________.

15. $C_6H_5—CH_2—OH$ is classified as a ______________________ alcohol.
(primary, etc.)

SUBSTITUTIVE NAMES

The rules learned for alkenes are followed for alcohols. The longest continuous chain of carbon atoms containing the functional group (OH in this case) serves as the basis of the name. The OH group is assigned the smaller possible locant (position number). The ending that signifies alcohol or OH group is ol. This systematic ending replaces the final e in the name of the alkane corresponding to the parent chain, and a locant for the OH group precedes the stem of the name. For example, the IUPAC substitutive name for

$$\begin{array}{c} CH_3-\underset{\displaystyle OH}{\underset{|}{CH}}-CH_3 \end{array}$$

is 2-propanol. Note that the name isopropanol for this (or any) compound is incorrect, because there is no parent alkane, isopropane.

16. The IUPAC substitutive name for

$$CH_3-CH_2-\underset{\displaystyle OH}{\underset{|}{CH}}-CH_2-CH_3$$

is ______________________.

17. The condensed structural formula for 2-butanol is

______________________________.

18. The systematic name for

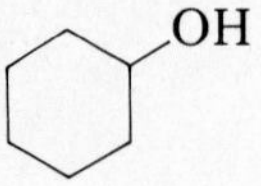

OH

is ______________________________.

Substituents are named and numbered as in other IUPAC names. Remember that the OH group is assigned the lower possible locant.

19. The substitutive name for

$$CH_3-\underset{\displaystyle OH}{\underset{|}{CH}}-\underset{\displaystyle CH_3}{\underset{|}{CH}}-CH_3$$

is ______________________________.

A complex structural formula can usually be named rather easily in steps. Items 20 through 26 apply to the formula

$$
\begin{array}{ccccccccccccccccc}
 & & & & \text{Cl} & & \text{C}_6\text{H}_5 & & & & & & & & & & \\
 & & & & | & & | & & & & & & & & & & \\
\text{CH}_3 & - & \text{CH}_2 & - & \text{CH} & - & \text{C} & - & \text{CH}_2 & - & \text{CH} & - & \text{CH}_2 & - & \text{CH}_2 & - & \text{CH}_3 \\
 & & & & & & | & & & & | & & & & & & \\
 & & & & & & \text{OH} & & & & \text{CH} & - & \text{CH}_3 & & & & \\
 & & & & & & & & & & | & & & & & & \\
 & & & & & & & & & & \text{CH}_3 & & & & & &
\end{array}
$$

and develop the name for it in steps.

20. The longest continuous chain of carbon atoms containing the functional group that will serve as the basis of the name contains ________ (number) carbon atoms. The IUPAC name for an alkane containing this number of carbon atoms is ____________________.

21. The IUPAC substitutive (parent) name for an alcohol with this many carbon atoms is

____________________.

22. When the parent chain is properly numbered, the OH group will be on the carbon atom numbered ____. Therefore, the parent compound name, including locant, will be

____________________.

23. There are ________ (number) substituents on the parent chain, besides the functional group; these substituents are named ____________, ____________, and ____________; they are located on carbons numbered ______, ______, and ______, respectively.

24. As in all IUPAC names, the locants are separated from the letter parts of the name by __________ (and from each other, if necessary, by __________).

25. The complete IUPAC name for the complex formula above then becomes

____.

26. This compound can be classified as a __________ alcohol.

27. Phenylmethyl alcohol, represented by the formula ______________________________, may also be given the substitutive name ______________________________________.
For indexing convenience, *Chemical Abstracts* uses for this compound the name

benzenemethanol—an example of another IUPAC style of nomenclature, conjunctive names. Even if one does not find this style of name in general use, there is little or no difficulty in translating such a name in *Chemical Abstracts* into the intended structure.

ALCOHOLS WITH TWO OR MORE FUNCTIONAL GROUPS

Some alcohols contain two or more OH groups. Substitutive names for these compounds simply include a locant for each OH group and the appropriate ending diol, triol, tetrol, etc. For example, $HO{-}CH_2{-}CH_2{-}OH$ may be named 1,2-ethanediol. Note that the final "e" of the alkane portion of the name is retained for alkanediol but is omitted for alkanol. In general, the final "e" is retained with systematic endings beginning with consonants and dropped with those beginning with vowels.

Some diols of low molecular weight are often named by two-word functional class names that use the class designation glycol, together with the bivalent hydrocarbon group name. For example, $HO{-}CH_2{-}CH_2{-}OH$ is commonly called ethylene glycol.

28. 1,2-Propanediol may be represented by the formula

______________________________.

and also identified by the functional class name, propylene glycol.

29. Its isomer, $HO{-}CH_2{-}CH_2{-}CH_2{-}OH$, has the substitutive name,

____________, and the functional class name, trimethylene glycol. In general, substitutive names for diols are preferred over the glycol names.

30. The substitutive name for the compound

$$\begin{array}{c} HO{-}CH_2{-}\underset{\displaystyle OH}{\underset{|}{CH}}{-}CH_2{-}OH \end{array}$$

is ______________________________. This compound is commonly called glycerol.

31. The parent chain in the formula

$$CH_3{-}\overset{\displaystyle OH}{\overset{|}{CH}}{-}CH_2{-}\overset{\displaystyle OH}{\overset{|}{CH}}{-}\overset{\displaystyle CH_3}{\overset{|}{CH}}{-}\underset{\displaystyle CH_3}{\underset{|}{\overset{\displaystyle CH_3}{\overset{|}{C}}}}{-}C_6H_5$$

is numbered so that the OH groups are assigned positions _____ and _____. The rule dictating the direction of numbering requires that the OH groups be assigned the

____________________ locants. The IUPAC name for the alcohol represented by this formula is ______________________________________.

32. The name for the compound

(cyclohexane ring with OH, OH, and HO substituents)

is ______________________________________.

Whenever a compound contains more than one functional group, one of the groups is chosen as the principal group and is the basis of the name of the compound (that is, the ending of the name). IUPAC rules include an order of priority for principal groups. (A partial list is in the Appendix.) All other functional groups outrank carbon-carbon multiple bonds.

IUPAC substitutive names of unsaturated alcohols are formed by using alkenol or alkynol rather than alkanol as the basis of the name. The position of the OH group is assigned the lower possible number. In cycloalkenols, the OH group is always on the number 1 carbon. No locant is necessary for it in the name, but the locant 1 is usually included anyway. The locant for the carbon-carbon multiple bond generally precedes the parent stem of the name, and the locant for the OH group immediately precedes ol. For example,

$$CH_3-CH=CH-\underset{\displaystyle OH}{\underset{|}{CH}}-CH_3$$

is named 3-penten-2-ol.

33. The structural formula for allyl alcohol is

____________________,

and the substitutive name for allyl alcohol is ____________________.

34. A structural formula for 2-cyclohexen-1-ol is

____________________.

35. A substitutive name for

—OH

is ______________________________.

36. Oblivon, a hypnotic, is 3-methyl-1-pentyn-3-ol, which may be represented by the formula

______________________________________.

37. The chain of carbon atoms that will serve as the basis of the IUPAC substitutive name of

```
          CH3               OH
          |                 |
   CH3—C=CH—CH—CH—CH2—CH3
                  |
                  CH2—CH3
```

contains ________ (number) carbons, and the portion of the name used to designate that chain together with the functional groups is ______________. When the carbon chain is properly numbered, the OH group will be on carbon numbered ______, and the alkene linkage will be assigned position number ______. The parent name then becomes ______________ ______________. There are ________ (number) substituents (not including the OH group) on the parent chain. They are named __________ and __________ and are located on carbons numbered ____ and ____, respectively. The complete IUPAC substitutive name for the unsaturated alcohol becomes ______________________________________.

38. A structural formula for 5-chloro-3-phenyl-3-hexen-2-ol is

______________________________________.

39. A substitutive name for

$$\underset{\displaystyle OH}{CH_3-\underset{|}{CH}}-CH=CH-\underset{\displaystyle OH}{\underset{|}{CH}}-\underset{\displaystyle CH_3}{\underset{|}{CH}}-CH_3$$

is ______________________________.

40. A substitutive name for $CH_3-CH_2-\overset{\displaystyle CH_3}{\overset{|}{CH}}-$ (ring bearing OH and OH) $-OH$ is

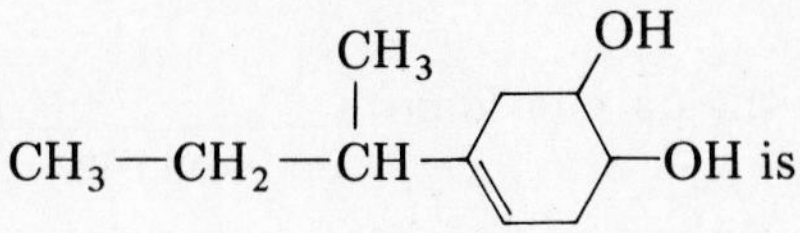

____.

Alkenols, like other alkenes, may exist as cis,trans isomers (see Chapter 3, items 31 through 46). The prefix *cis*- or *trans*- precedes the locant(s) for the functional group(s) to which the geometrical tag applies.

41. A structural formula for 2-methyl-*trans*-3-penten-2-ol is

______________________________.

42. A completely descriptive name, including identification of configuration (spatial arrangement), for the compound represented by the formula

$$HO-CH_2-\underset{\displaystyle CH_3}{\underset{|}{CH}}-\underset{\displaystyle H}{\underset{|}{C}}=\underset{\displaystyle H}{\underset{|}{C}}-CH_2-\underset{\displaystyle C_6H_5}{\underset{|}{CH}}-CH_2-CH_2-CH_2-CH_3$$

is____.

43. By use of a geometric figure for the ring structure, *cis*-2-*trans*-6-cyclodecadien-1-ol may be represented by the structure

______________________________.

44. Even so complex an alcohol as vitamin A,

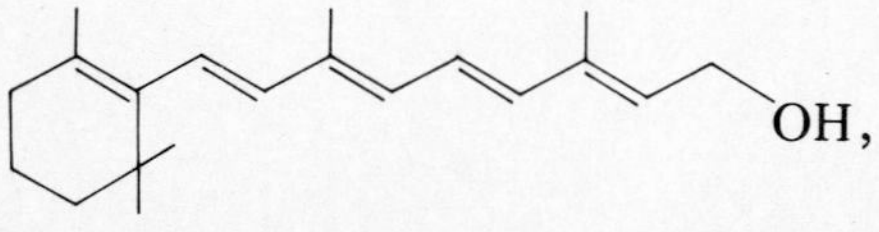

OH,

can be properly named rather easily by a substitutive name. The cyclic and acyclic portions of the line drawing shown have the same significance; that is, at each intersection or termination point, there is a carbon and as many hydrogens as necessary to complete a valence of 4. The ring portion of the compound is a substituent on the last of a parent chain of ________ carbons. That substituent is named ______________________________.
(number)

Each of the alkene linkages in the parent chain is ______________. The full name of the
(cis or trans)

substituted, unsaturated alcohol illustrated is

____.

OH AS SUBSTITUENT

If an OH group is not attached to the parent chain (or is for other reasons to be treated as a substituent rather than the basis of the name), the prefix hydroxy is used to signify an OH substituent. This substituent name is used just as any other substituent name.

45. The compound whose substitutive name is 2-(hydroxymethyl)-1,3-propanediol has the formula

____________________________________.

46. The formula

$$
\begin{array}{lllll}
 & CH_3 & & & \\
 & | & & & \\
 & CH_2 & & OH & \\
 & | & & | & \\
HO-C_6H_4- & CH-CH_2-CH_2- & & CH-CH_3 &
\end{array}
$$

shows on position 5 a substituent named ______________________; the name of the compound is ____.

7 Ethers

Compounds having the structure represented by the general formula $R-O-R'$ are classified as ethers. R and R' may be the same groups (symmetrical ethers) or different groups (unsymmetrical ethers). Substitutive names are preferred for ethers, but multiple-word, functional class names are still used, especially for some familiar, symmetrical ethers. *Chemical Abstracts* does not use the functional class names anymore in indexes.

SUBSTITUTIVE NAMES

There is no systematic ending for the ether linkage in substitutive names. Ethers are named by considering $R-O-$ as a substituent that replaces H in the parent compound, $H-R'$. The name of the $R-O-$ substituent is formed by combining the name for the group R with oxy; for example, $CH_3-CH_2-CH_2-CH_2-CH_2-O-$ is pentyloxy. A shortened name, illustrated by the general name alkoxy, is used when R is an alkyl group containing fewer than five carbons or is phenyl; for example, CH_3-CH_2-O- is ethoxy and C_6H_5-O- is phenoxy. No particular preference is given to an oxy substituent over other substituents in assigning locants on the parent chain. Methoxymethane is the substitutive name for the symmetrical ether, CH_3-O-CH_3.

1. The substitutive name of the ether $CH_3-CH_2-O-CH_2-CH_3$ is

______________________________.

2. The longest continuous chain of carbons in the formula

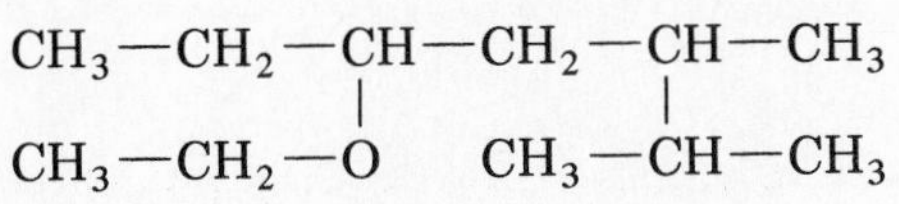

$$
\begin{array}{ccccccccccc}
CH_3 & - & CH_2 & - & CH & - & CH_2 & - & CH & - & CH_3 \\
 & & & & | & & & & | & & \\
CH_3 & - & CH_2 & - & O & & CH_3 & - & CH & - & CH_3
\end{array}
$$

contains __________ carbons. The name for this chain is ______________________________.
(number)

Substituents appear on carbons numbered ____________________. Two of the substituents are called methyl, and the third is called __________. A complete substitutive name for this ether is ____.

3. A structural formula for 1,4-dimethoxybenzene is

________________________________,

and one for 1,1-diethoxybutane is ________________________________.

4. The substitutive name for the ether $C_6H_5—O—C_6H_5$ is ________________________________.

5. The name of

$$\text{C}_6\text{H}_{11}\text{—O—}\underset{}{\overset{\text{CH}_3}{\overset{|}{\text{CH}}}}\text{—CH}_2\text{—}\overset{\text{Cl}}{\overset{|}{\text{CH}}}\text{—CH}_2\text{—}\underset{\underset{\text{CH}_3}{|}}{\overset{\text{CH}_3}{\overset{|}{\text{C}}}}\text{—CH}_3$$

will be based on the parent compound, ________________________________. There are

________ substituents on this parent chain at positions numbered ________________________.
(number)

The complete name for the ether is ________________________________.

6. A structural formula for 1,4-dimethoxy-<u>trans</u>-2-pentene is

________________________________.

7. A substitutive name for the alcohol

$$\text{CH}_3\text{—CH}_2\text{—O—}(\text{cyclohexane ring with OH})\text{—C}_6\text{H}_5$$

will use ________________________ as the parent name and will specify two substituents,

named ________________ and ________________. Correct numbering will assign the OH group to

position number ________ and the ethoxy substituent to position number ________. The

complete substitutive name for this alcohol is

____.

8. A substitutive name for the alkyne

$$
\begin{array}{lllll}
 & & & CH_3 & \\
 & & & | & \\
CH_3-C\equiv C-CH_2- & CH & -O- & CH & -CH_3 \\
 & | & & & \\
 & \rlap{CH_2-CH_2-CH_2-CH_2-CH_3} & & &
\end{array}
$$

will be based on the longest continuous chain of carbons, containing ________ carbons.
(number)

The alkyne name for this chain is ____________. The functional group that is the basis of the name is assigned position number ________, and the substituent, named ______________, is on position number ________. The complete substitutive name for this compound is ______________________________________.

FUNCTIONAL CLASS NAMES

Ether is a class name, not the name of any individual member of the class. Names ending in ether are therefore multiple-word names, not single-word names, and are formed by preceding ether with the name of each group attached to oxygen. For example, $CH_3-O-CH_2-CH_3$ may be named ethyl methyl ether, and CH_3-O-CH_3, dimethyl ether. Although IUPAC accepts functional class names for unsymmetrical ethers, current recommendations limit these names to symmetrical ethers.

9. Diphenyl ether may be represented by the formula

______________________________________,

and isopropyl phenyl ether by the formula

______________________________________.

10. The compound represented by the formula $CH_2=CH-CH_2-O-CH_2-CH=CH_2$ shows attached to oxygen two groups that have the IUPAC common (trivial) name ______________, and the compound represented by the formula has the functional class name ______________________________________.

11. Di-sec-butyl ether may be named with the substitutive name,

______________________________.

12. The trivial name for the group

$$\begin{array}{c} \quad\quad\quad\quad\quad\quad CH_3 \\ \quad\quad\quad\quad\quad\quad | \\ CH_3-CH_2-C- \\ \quad\quad\quad\quad\quad\quad | \\ \quad\quad\quad\quad\quad\quad CH_3 \end{array}$$

is ______________________________, the name of the group ▷— is ______________

______________, and the functional class name for the compound

$$\begin{array}{c} \quad\quad\quad\quad\quad\quad CH_3 \\ \quad\quad\quad\quad\quad\quad | \\ CH_3-CH_2-C-O-\triangleleft \\ \quad\quad\quad\quad\quad\quad | \\ \quad\quad\quad\quad\quad\quad CH_3 \end{array}$$

is ______________________________. The preferred, substitutive

name for this compound is ______________________________.

POLYETHERS

Compounds containing more than two ether linkages as part of a continuous chain of atoms are most easily named by a different style of IUPAC nomenclature, replacement names. The continuous chain is considered a parent chain, and the oxygen atoms are considered to be replacements for carbons in that chain. A multiplying prefix (di, tri, etc.) and the prefix oxa signify replacement of carbons in the parent chain by oxygens, and locants preceding the prefixes designate the location along the chain of those replacing atoms. For example, $CH_3-O-CH_2-CH_2-O-CH_2-CH_2-O-CH_3$, a useful solvent nicknamed diglyme, may be properly named 2,5,8-trioxanonane. This replacement name, which tells us that oxygens replace carbons at positions 2, 5, and 8 in a nonane (nine-carbon) chain, is more convenient and is shorter than the alternative, substitutive name 1-methoxy-2-(2-methoxyethoxy)ethane. Although replacement names may be correctly used for compounds with only one or two ether linkages, they offer little or no advantage over the substitutive names, which are favored for these compounds. Replacement names are intended to be used for compounds difficult to name in other approved ways.

Note that the prefix, oxy signifies a substituent on a parent chain, and the prefix oxa signifies a replacement of a carbon in the parent chain.

13. 1,3,5-Trioxacyclohexane may be represented by the formula

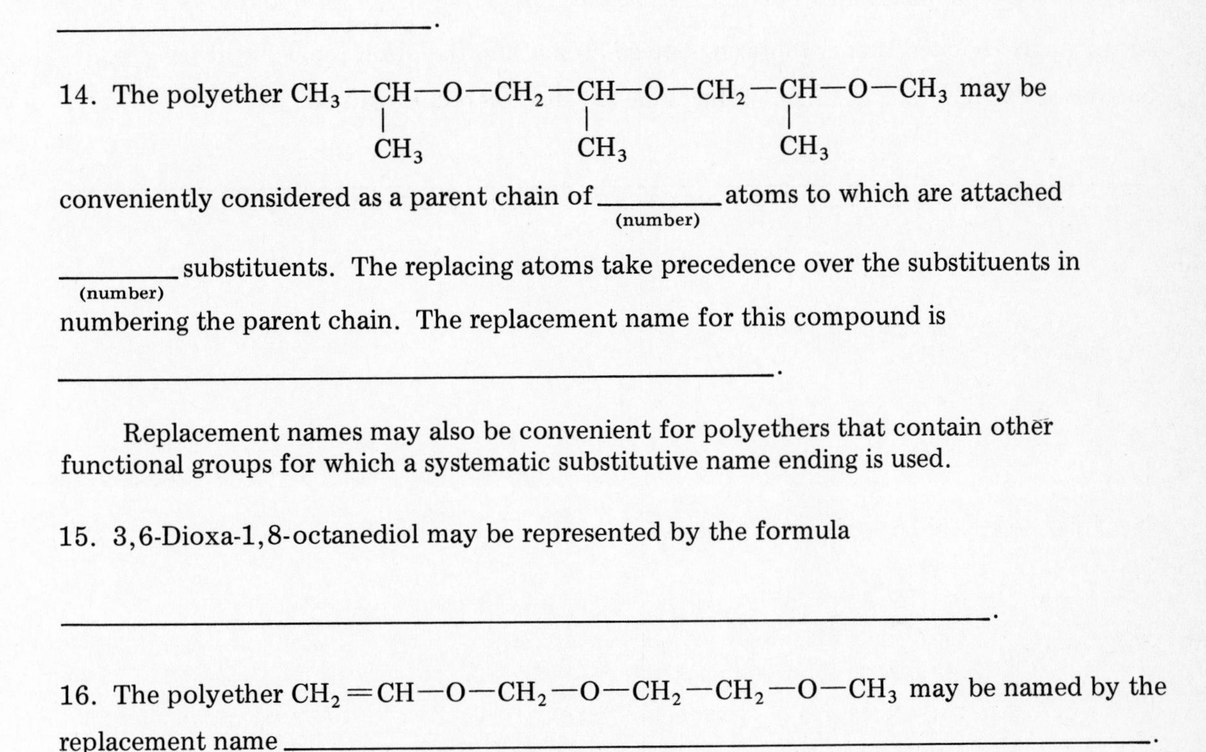

____________________________.

14. The polyether $CH_3-\underset{\displaystyle CH_3}{\underset{|}{CH}}-O-CH_2-\underset{\displaystyle CH_3}{\underset{|}{CH}}-O-CH_2-\underset{\displaystyle CH_3}{\underset{|}{CH}}-O-CH_3$ may be conveniently considered as a parent chain of ________ (number) atoms to which are attached ________ (number) substituents. The replacing atoms take precedence over the substituents in numbering the parent chain. The replacement name for this compound is

____.

Replacement names may also be convenient for polyethers that contain other functional groups for which a systematic substitutive name ending is used.

15. 3,6-Dioxa-1,8-octanediol may be represented by the formula

____.

16. The polyether $CH_2{=}CH-O-CH_2-O-CH_2-CH_2-O-CH_3$ may be named by the replacement name ____.

Atoms replacing carbons in a parent compound are called hetero atoms, and replacement names are often convenient for compounds that illustrate multiple replacements of carbon by hetero atoms other than oxygen. A different prefix (ending in a) is used for each different replacing atom (aza for N, thia for S, sila for Si), and the citation sequence for those prefixes, rather than being strictly alphabetical, always begins with oxa (if present). (See the Appendix for priority sequence for these prefixes.) For example, $CH_3-O-CH_2-CH_2-NH-CH_2-CH_2-O-CH_3$ may be considered as a chain of nine atoms (with three hetero atoms in the chain) and named 2,8-dioxa-5-azanonane. The use of replacement names is likely to grow with use of computer-based literature searches, because the names are convenient for indexing.

17. A convenient, replacement name from the compound

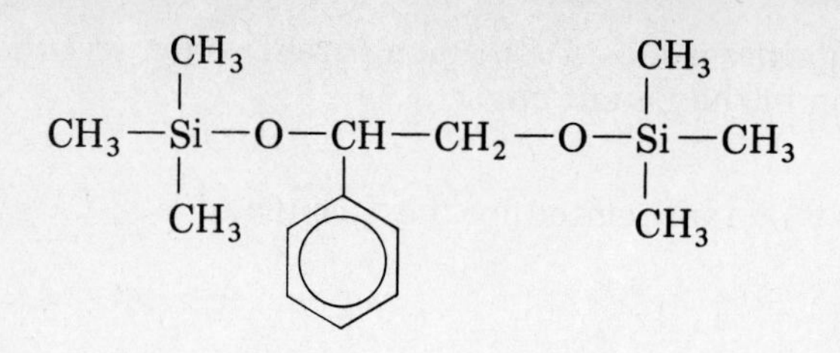

will be based on a parent chain of ________ (number) atoms, including ________ (number) hetero atoms. The oxygens are cited first as replacing hetero atoms, and the other replacing atoms are designated by the prefix _______. The name for the compound will be

____.

8 Substitution Products from Aromatic Hydrocarbons

Just as with aliphatic hydrocarbons, hydrogens in aromatic hydrocarbons may be replaced by a variety of substituent groups. Substituted benzenes (arenes) are named in essentially the same way as are substituted alkanes. For example, C_6H_5—Cl is named chlorobenzene. Some other common substituents, in both alkanes and arenes, are: —F, fluoro; —Br, bromo; —I, iodo; —NO_2, nitro; —NO, nitroso; and —CN, cyano.

1. The substituted benzene represented by the formula

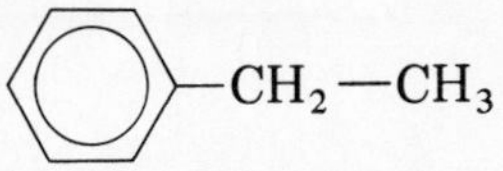

C_6H_5—CH_2—CH_3

is named ____________________; that represented by the formula

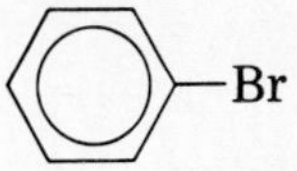

C_6H_5—Br

is named ____________________; that represented by the formula

C_6H_5—NO_2

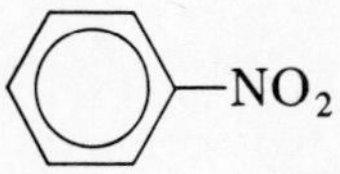

is named ____________________; and that represented by the formula

C_6H_5—NO

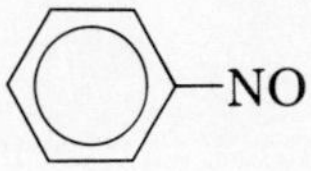

is named ____________________.

The relative positions of substituents in multiply substituted benzenes are indicated by letters (o-, m-, or p- for disubstituted compounds) or by numbers. When numbers are used in the name of a substituted benzene, any substituent may be assigned position number 1, with the limitation that the smallest possible locants for substituents must be used. For simplicity of indexing, the substituents are cited in alphabetical order.

2. Formulas for three isomeric dichlorobenzenes may be drawn

________________, ________________, and ________________; the compounds represented by these formulas may be named, with letters used to designate relative positions of substitution, ________________, ________________, and ________________, respectively, or, with numerical locants, ________________, ________________, and ________________, respectively.

3. m-Chloronitrobenzene may be represented by the formula

________________.

If numbers are used in the name, m-chloronitrobenzene may be named ________________.

4. Picryl chloride is a common name for the very reactive compound

Cl

O_2N NO_2

NO_2

The smallest numerical locants that can be used in a systematic name for picryl chloride are ____, ____, ____, and ____. A systematic name for this substituted benzene, with numerical locants being used to indicate positions of substitution, is ________________________________.

5. 2,4,6-Trinitrotoluene, commonly known as TNT, may be represented by the formula

________________________.

6. Even though the formulas for picryl chloride (item 4) and TNT (item 5) closely resemble each other, the locants for the nitro substituents are different when the compounds are named as derivatives of different hydrocarbons. Different locants are used

because ____

____.

TNT (item 5) may also be named as a substituted benzene; its name then is

____. Compare the locants in this name with those in item 4.

In compounds related to alkylbenzenes, a substituent may be bound to a carbon in the benzene nucleus or to a carbon in the alkyl group. The alkyl group is called a side chain.

7. Formulas may be drawn to represent three isomeric compounds formed by replacing a hydrogen in the aromatic nucleus of toluene by a chlorine. The three formulas are

________________________, ________________________, and ________________________,

and the three substituted toluenes may be named ____________________________,

______________________________, and ______________________________, respectively.

8. A fourth isomer, in which the chloro substituent appears on the side chain rather than in the benzene nucleus, may be represented by the formula,

________________________. When the substituent appears on the side chain of toluene, the position of substitution may be designated by the Greek letter alpha (α), but this usage is not recommended. The entire side chain should be named as a substituent on the parent, benzene.

9. Toluene might be considered as the parent compound in a name for

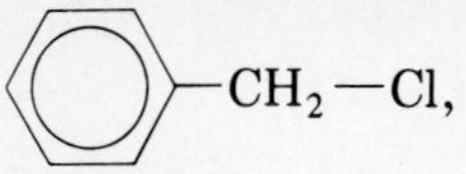

$C_6H_5-CH_2-Cl$,

but the compound is better named as a substituted benzene. The single substituent is named ____________________, and the substitutive name for the compound is ____________________.

10. The compound by the formula

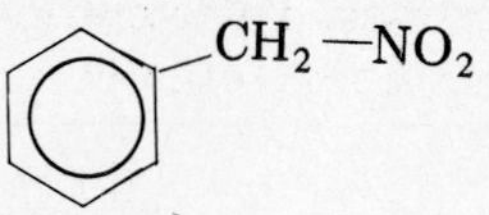

$C_6H_5-CH_2-NO_2$

is named ____________________.
Parentheses around the names of the substituents are needed in items 9 and 10 because

____

____.

For purposes of nomenclature, the group

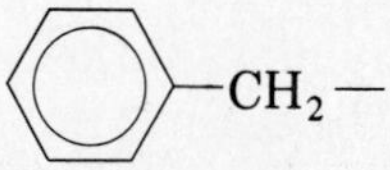

$C_6H_5-CH_2-$

is often regarded as an alkyl group and is named benzyl. The name benzyl is used in the same way as the name methyl.

11. A functional class name for the compound represented by the formula

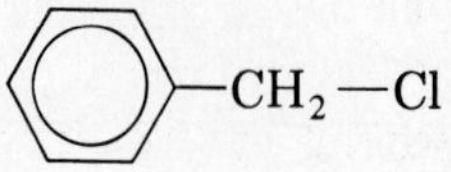

$C_6H_5-CH_2-Cl$

is ________________________________; a functional class name for the compound

$C_6H_5-CH_2-I$

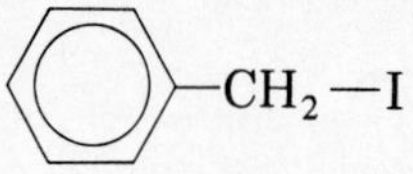

is ________________________________; and a functional class name for the compound

$C_6H_5-CH_2-O-CH_2-C_6H_5$

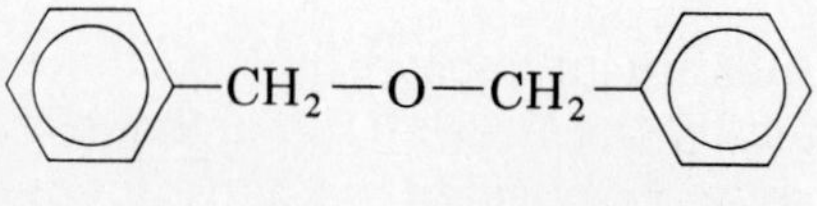

is ________________________________.

12. Benzyl alcohol, represented by the formula

______________________________,

is classified as a ____________ alcohol.
(primary, etc.)

Benzyl is accepted by IUPAC, but its use for the unsubstituted group only is recommended. *Chemical Abstracts* uses phenylmethyl instead of benzyl for this group.

13. The compounds

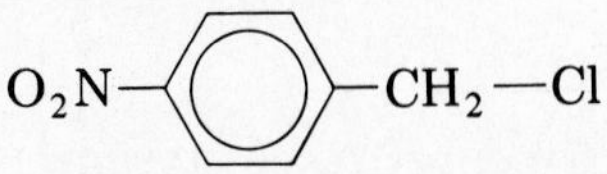

O_2N— —CH_2—Cl and CH_2—OH OCH_3

A B

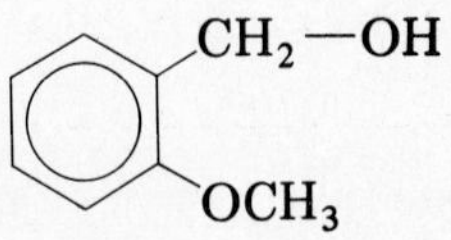

contain substituted benzyl groups, so benzyl names are not recommended. Compound A is preferably named as a disubstituted benzene,

____, and compound B is preferably
(compound name)

named by a substitutive name for an alcohol, ____.
(compound name)

Phenyl and benzyl are names that are sometimes confused by beginning students.

14. Phenylmagnesium bromide is a Grignard reagent represented by the formula

______________________,

and benzylmagnesium bromide is represented by the formula

______________________.

15. Benzyl bromide,

______________________________,
(formula)

and p-bromotoluene,

______________________________,
(formula)

are isomers.

16. Recall that styrene may serve as a parent name for substituted styrenes. Styrene, an important industrial chemical, is represented by the formula

______________________________.

17. 4-Nitrostyrene may be represented by the formula

______________________________;

it will be indexed in *Chemical Abstracts* as a disubstituted benzene,

____.
(compound name)

18. 2,4-Difluorostyrene may be represented by the formula

______________________________.

19. If the compound represented by the formula

$$C_6H_5-CH_2-C_6H_3(F)-CH=CH_2$$

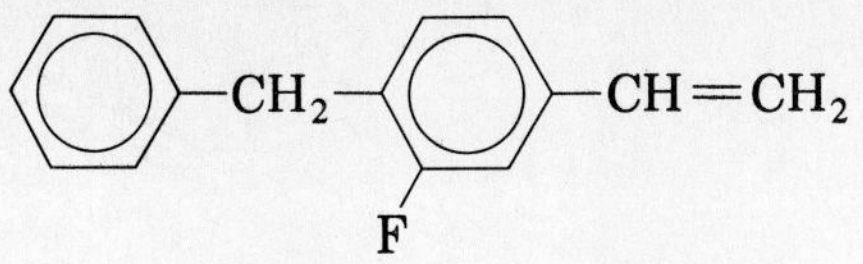

is named as a substituted styrene, the substituent in the 4-position will be named

________________, and the name of the compound will be ____________________________________.

20. Naphthalene is represented by the formula

______________________,

and 1,5-dibromonaphthalene is represented by the formula

______________________.

21. [naphthalene ring with CH_3, NO_2, and NO_2 substituents]

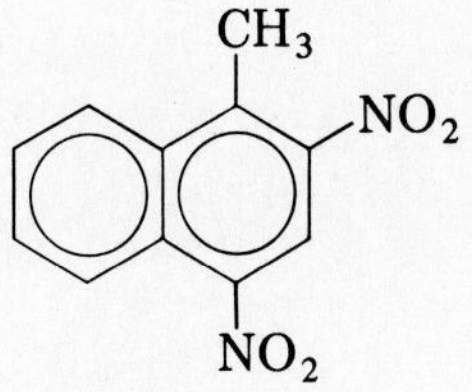

is named ____.

22. [tetrahydronaphthalene ring with $O-CH_2-CH_3$ substituent]

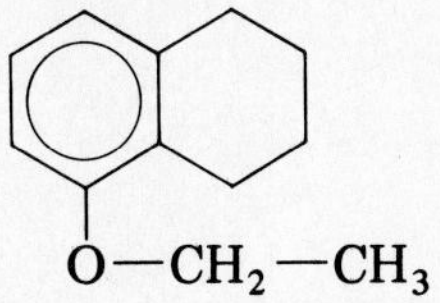

may be named as a naphthalene; on that basis, hydro will be used to designate

____,

and the name of the compound will be ____.

23. Biphenyl is represented by the formula

______________________________,

and 2,4′-dibromo-4-nitrosobiphenyl is represented by the formula

______________________________.

9 Acids

CARBOXYLIC ACIDS

The functional group $-\overset{\overset{\large O}{\|}}{C}-OH$ (or $-COOH$ or $-CO_2H$) is called a carboxy group, and compounds containing this functional group are carboxylic acids. Several types of IUPAC-approved names are used for carboxylic acids; structural features of the compound other than the carboxy group often determine the type of name most frequently used.

SUBSTITUTIVE NAMES (ACYCLIC CARBOXYLIC ACIDS)

If one imagines that a terminal CH_3 group of an acyclic hydrocarbon (alkane, alkene, or alkyne) is transformed into a COOH group, the acid is named by combining the hydrocarbon name (minus the final e) with the systematic ending, oic acid. The stem of the acid name indicates the number of carbons in the parent compound, including the one in the carboxy group. For example, CH_3-CH_2-COOH is propanoic acid.

In such names, the C in the COOH group is always position number 1, but the locant is not included in the name. This locant assignment takes precedence over substituents and other functional groups in -oic acid names.

1. The structural formula for butanoic acid is ________________________________,

and that for 2-butenoic acid is ________________________________.

2. The substitutive name for the acid

$$\begin{array}{l} CH_3-\underset{\displaystyle |}{C}H-CH_2-COOH \\ \qquad\;\; CH_2-CH_2-CH_3 \end{array}$$

will be based on a parent chain of __________ (number) carbons, for which the stem is ______________.

A substituent named ________________ is at position number __________. The name of the acid is ____.

3. The compound $CH_2{=}CH{-}CH_2{-}CH_2{-}CH_2{-}CH_2{-}CH_2{-}CH_2{-}CH_2{-}COOH$ is named ____.

4. The parent chain in the formula

$$
\begin{array}{c}
\qquad\qquad\qquad\quad C_6H_5{-}CH_2 \\
\qquad\qquad\qquad\qquad\quad | \\
CH_3{-}CH_2{-}\underset{\displaystyle CH_3}{\underset{|}{C}H}{-}CH_2{-}\underset{\displaystyle NO_2}{\underset{|}{C}}{-}CH_2{-}COOH
\end{array}
$$

contains ________ (number) carbons. Substituents appear on carbons numbered ______________.

The name for this acid is ____.

5. Substituents named ethynyl and ethoxy are represented by the formulas ______________ and ______________, respectively. A structural formula for 5-ethoxy-2-ethynylpentanoic acid is

____.

6. Naphthalene has the formula

______________________________,

and 2-naphthyloxy is a substituent group with the formula

______________________________.

2-(2-Naphthyloxy)ethanoic acid has the formula

______________________________.

7. $C_6H_5-C\equiv C-CH_2-CH_2-COOH$ will be named ______________________________.

8. The parent compound on which the name of

$$
\begin{array}{l}
\qquad\qquad\quad CH_2-CH_3 \\
\qquad\qquad\quad | \\
Br-C_6H_3(Br)-CH-CH-CH_2-COOH \\
\qquad\qquad\qquad\quad | \\
\qquad\qquad\qquad\quad O-CH_2-CH_3
\end{array}
$$

will be based is named ______________________________, and the substituents are named ______________________________ ______________________________. The name of the compound illustrated is ______________________________.

9. Because of the geometry of some unsaturated linkages, cis,trans isomers are possible for ______________ acids but not for ______________.
(alkenoic or alkynoic) (alkenoic or alkynoic)

10. In the name 2-ethyl-trans-3-pentenoic acid, trans refers to ______________________________ ______________________________, and a structural formula for the compound is

______________________________. A structural formula for the isomer, 2-ethyl-cis-3-pentenoic acid, is

______________________________.

Acyclic compounds containing two carboxy groups are named by combining the ending -dioic acid with the name of the corresponding hydrocarbon. For example, HOOC—COOH may be named ethanedioic acid. Note that the treatment of the final e of the hydrocarbon name parallels that in the substitutive names of alcohols: the e is dropped for addition of a vowel ending (-ol and -oic acid) but is retained for addition of a consonant ending (-diol and -dioic acid).

11. Hexanedioic acid contains ________ carbons and is represented by the formula
(number)

____.

12. The compound

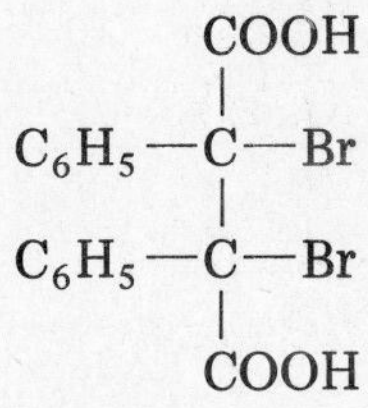

$$\begin{array}{c} COOH \\ | \\ C_6H_5-C-Br \\ | \\ C_6H_5-C-Br \\ | \\ COOH \end{array}$$

may be named ____.

13. The compound

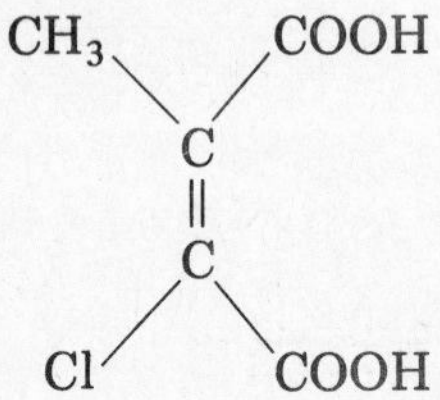

$$\begin{array}{ccc} CH_3 & & COOH \\ & \diagdown \; \diagup & \\ & C & \\ & \| & \\ & C & \\ & \diagup \; \diagdown & \\ Cl & & COOH \end{array}$$

is shown in the ______ configuration, and its name will be based on the parent compound
(E, Z)

(acid) ________________________________. The name of the compound illustrated

is ____.
(include E, Z)

14. (E)-3,7-Dimethyl-2-octenedioic acid is a component of a bean weevil copulation release hormone, nicknamed erectin; the structural formula for this compound is

________________________.

SUBSTITUTIVE NAMES (CYCLIC CARBOXYLIC ACIDS)

Compounds in which a carboxy group is attached to a ring system are named by combining the name of the ring system with the ending -carboxylic acid. For example, ▷—COOH is cyclopropanecarboxylic acid. Note that the stem of the name does not include the carbon in the carboxy group ("carboxylic" counts a carbon) and that the final e is retained for combination with a consonant ending. Except in rings with fixed numbering (naphthalene, for example), the carbon to which the carboxy group is attached is the number 1 position, and all other locants follow from that one. The locant 1 does not usually appear in the name. Names of compounds containing more than one carboxy group include the appropriate multiplier in the ending (for example, -dicarboxylic acid), and locants for all carboxy groups must be used for clarity.

15. 4-tert-Butylcyclohexanecarboxylic acid may be represented by the formula

______________________________________.

16. The isomers

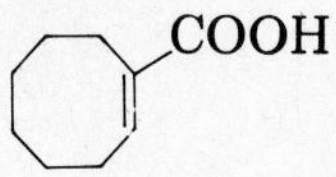

COOH and COOH

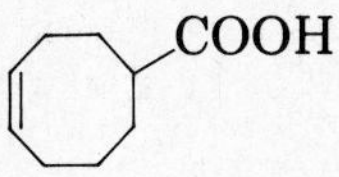

are named ______________________________________ and ______________________________

________________, respectively.

17. 1,2,4-Benzenetricarboxylic acid contains a total of ________ carbons; its structural
(number)

formula is

________________________________.

18. The naphthalene derivative,

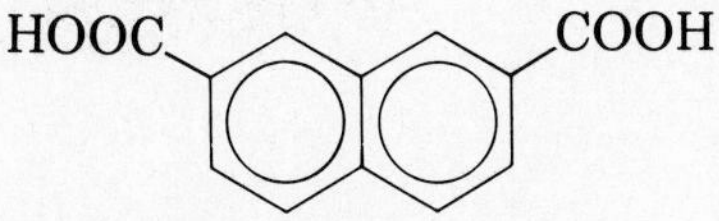

HOOC COOH

is named ____, and the biphenyl derivative,

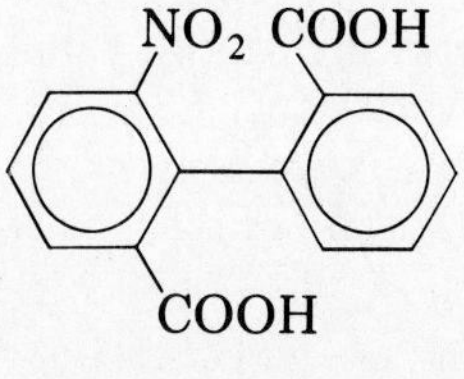

NO_2 COOH

COOH

is named ____.

19. 1,4-Dihydronaphthalene-2-carboxylic acid is represented by the formula

________________________________.

Acyclic compounds containing more than two carboxy groups are conveniently named in this style, since it permits the maximum number of the same functional group to be treated alike.

20. The name 1,1,3,3-propanetetracarboxylic acid indicates a total of ________ (number) carbons in the compound. The compound has the formula

____.

21. The compound

$$\begin{array}{ccccccc} HOOC-CH_2- & CH & -CH_2- & CH & -CH_2-COOH \\ & | & & | & \\ & COOH & & COOH & \end{array}$$

may be named with a name that treats all the carboxy groups alike; that name is ____. When all the carboxy groups <u>cannot</u> be treated alike in the name, some (the minimum number) are treated as substituents for which the prefix carboxy is used.

TRIVIAL NAMES (ARENECARBOXYLIC ACIDS)

Some arenecarboxylic acids are nearly always identified by shortened, trivial names.

22. Benzenecarboxylic acid, whose formula is

______________________,

is usually named benzoic acid, and the isomeric naphthalenecarboxylic acids,

______________________ (formula) and ______________________ (formula), are usually named naphthoic acids. IUPAC accepts these trivial names, as well as the longer, more systematic ones, and the trivial ones are used by most chemists. *Chemical Abstracts* uses benzoic acid for that simple acid but uses the longer, -carboxylic acid names for some benzoic acid derivatives and for other arenecarboxylic acids. Locants for substituents are used in the same way in both kinds of names.

23. 3, 5-Dinitrobenzoic acid may be represented by the formula

______________________________.

24. The formula

COOH

NO_2

shows a carboxy group attached to the ________ (number) position of naphthalene. The compound may be named by the longer IUPAC name used by *Chemical Abstracts* index,

____, or by the shorter IUPAC

trivial name used by many chemists, _____________________________________.

25. The name of the compound

$CH_3-C\equiv C-C_6H_4-COOH$

may be based on a parent acid with the (IUPAC) trivial name ________________________

________. The substituent is named __________________________, and the compound

illustrated is named ____.

TRIVIAL NAMES (ALKANOIC ACIDS)

Unsystematic names of some carboxylic acids were so well established by the time that rigid rules for systematic nomenclature were attempted by an International Congress in 1892 that they could not be discarded. These trivial names have been retained for some low-molecular weight carboxylic acids (those containing fewer than six carbons) and are still given preference by the IUPAC rules. *Chemical Abstracts* uses the trivial names only for the first two alkanoic acids, however, and the systematic names (alkanoic acids) are likely to become the more favored ones in general use for the other members of this class.

The trivial names of alkanoic acids are formed by adding to the proper stem the ending ic and the separate word acid. The stems used for carboxylic acids are quite different from those learned for alkyl groups. They are usually Latin or Greek in origin and often reflect the natural sources from which the acids were first isolated. The stems are associated with particular numbers of carbons and structures just as the stems for alkyl groups are.

Trivial names for some alkanoic acids are listed below. The stems, which will figure in names of acid derivatives, are underlined.

formic acid	$H{-}COOH$
acetic acid	$CH_3{-}COOH$
propionic acid	$CH_3{-}CH_2{-}COOH$
butyric acid	$CH_3{-}CH_2{-}CH_2{-}COOH$
isobutyric acid	$CH_3{-}\underset{\displaystyle CH_3}{\underset{\vert}{CH}}{-}COOH$
valeric acid	$CH_3{-}CH_2{-}CH_2{-}CH_2{-}COOH$
isovaleric acid	$CH_3{-}\underset{\displaystyle CH_3}{\underset{\vert}{CH}}{-}CH_2{-}COOH$

Only the stems of names signifying three and four carbons in carboxylic acids resemble the stems of names of alkyl groups. Locants for substituents are used in these trivial names as usual; the C of the carboxy group is position number 1.

26. The stem for three carbons in an alkyl group is ________, and that for three carbons in a carboxylic acid is ________. The stem for four carbons in an alkyl group is ________, and that for four carbons in a carboxylic acid is ________.

All other stems for the trivial names of carboxylic acid are completely different from the stems used for alkyl groups of corresponding carbon content.

27. The most common carboxylic acid contains two carbons and may be represented by the structural formula ____________. The trivial name for this acid is ________________.

28. The stem acet will always signify ________ (number) carbons in a carboxylic acid or acid derivative.

29. Since there is only one position available for a substituent in acetic acid, no locant is necessary. The trivial name for the substituted acid represented by the formula

$Cl—CH_2—COOH$ is ________________, and the trivial name for the acid represented by the formula $Cl_3C—COOH$ is ________________.

Note that chloroacetic and trichloroacetic are single words. Acetic acid is the name of an individual compound, and modifiers (substituent prefixes) must be written as part of the same word.

30. tert-Butoxy is the name of the substituent represented by the formula

________________,

and tert-butoxyacetic acid may be represented by the formula

________________.

31. Methanoic acid is the systematic name for the compound represented by the formula ________________, but the trivial name, ________________, is actually used more frequently than the systematic name for this compound.

PEROXY ACIDS

The —O—O— linkage is a peroxide linkage, and acids that contain —O—OH in place of —OH are called peroxy acids. (Note that peroxy acid is a two-word designation.) Peroxy acids are named by incorporating the unseparated prefix peroxy into the name of the corresponding acid. For acids named with -oic acid or -ic acid endings, peroxy precedes the stem portion of the name; for example

$$CH_3—\overset{\overset{\displaystyle O}{\|}}{C}—O—OH$$

is peroxyacetic acid. (For convenience, a peroxy acid may be represented by a formula such as $CH_3—CO_3H$.) For acids named with the -carboxylic acid ending, peroxy immediately precedes carboxylic.

32. The peroxy acid of lowest molecular weight is HCO_3H, which is named

________________.

33. m-Chloroperoxybenzoic acid has the formula

________________________________.

34. Cyclohexaneperoxycarboxylic acid has the formula

________________________________.

35. The compound $CF_3{-}CO_3H$ is named ________________________________.

A few frequently used peroxy acids have been identified by shortened names in which per replaces peroxy as the prefix. Although accepted for three peroxy acids by IUPAC, discontinuance of such names has been recommended. Consistent use of the prefix peroxy actually makes naming (and comprehension) easier.

SULFONIC ACIDS

Carboxylic acids are organic acids that contain a carboxy group, but other organic acids include the sulfonic acids, compounds that contain a sulfo group, $-SO_3H$, bound to carbon. Sulfonic acids are named in the same style as are cyclic carboxylic acids; that is, the ending -sulfonic acid is combined with the name of the parent hydrocarbon (final e retained), and the sulfo group is given the lowest possible locant. For example, $CH_3{-}CH_2{-}\underset{\displaystyle CH_3}{\underset{|}{CH}}{-}SO_3H$ is 2-butanesulfonic acid.

36. Benzenesulfonic acid has the formula ________________________________, and 4-methylbenzenesulfonic acid has the formula

________________________________.

4-Methylbenzenesulfonic acid is often called p-toluenesulfonic acid; that is, toluene rather than benzene is considered the parent hydrocarbon. *Chemical Abstracts* uses the benzene name, however, which is simpler in approach though longer.

37. The name of the compound

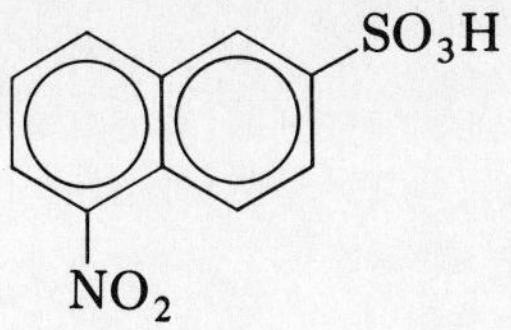

$$\text{(naphthalene ring with } SO_3H \text{ and } NO_2 \text{ substituents)}$$

will be based on the parent hydrocarbon, ______________________. The sulfo group will have the lowest possible locant, __________, and the compound is named

____.

38. The compound $CF_3{-}SO_3H$ is named ______________________________________.

39. 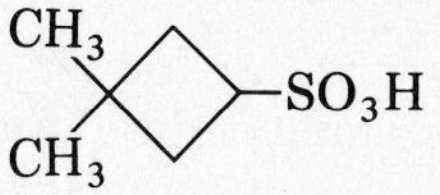 (CH_3, CH_3 on cyclobutane)$-SO_3H$ is named ______________________________________.

10
Acid Derivatives

ACID ANHYDRIDES

Removal of an OH group from a carboxylic acid generates an acyl group ($R{-}\overset{\overset{\displaystyle O}{\|}}{C}{-}$ or $R{-}CO{-}$). If two acyl groups are attached to oxygen ($R{-}\overset{\overset{\displaystyle O}{\|}}{C}{-}O{-}\overset{\overset{\displaystyle O}{\|}}{C}{-}R$, or for convenience in typing, $R{-}CO{-}O{-}CO{-}R$), the compound is an acid anhydride. An acid anhydride is related to an acid by the loss of HOH between two molecules of acid. If R and R′ are the same, the compound is a symmetrical acid anhydride; if different, an unsymmetrical acid anhydride.

Symmetrical acid anhydrides are named simply by replacing acid in the name of the corresponding acid with anhydride.

1. The trivial name of the most common carboxylic acid, $CH_3{-}COOH$, is ______________________, and the name of the related compound, $CH_3{-}CO{-}O{-}CO{-}CH_3$, is ______________________.

2. Benzoic acid has the formula ______________________, and benzoic anhydride has the formula ______________________.

3. Trifluoroacetic anhydride has the formula ______________________.

4. The acid

(cyclohexane ring)—COOH

is named ______________________, and the acid anhydride

(cyclohexane ring)—CO—O—CO—(cyclohexane ring)

is named ______________________.

The same type of nomenclature is used for acid anhydrides related to acids other than carboxylic acids.

5. The strong acid CF_3—SO_3H is named ______________________________,
and the acid anhydride CF_3—SO_2—O—SO_2—CF_3 is named

____.

Cyclic anhydrides of dicarboxylic acids are well known and are named in the same way as are acyclic anhydrides.

6. 1,2-Benzenedicarboxylic acid has the formula

____________________,

and the cyclic anhydride related to it, 1,2-benzenedicarboxylic anhydride, has the formula

____________________.

7. Butanedioic anhydride may be represented by the formula

____________________. This compound is usually identified by its trivial name, succinic anhydride.

8. The acid anhydride represented by the formula

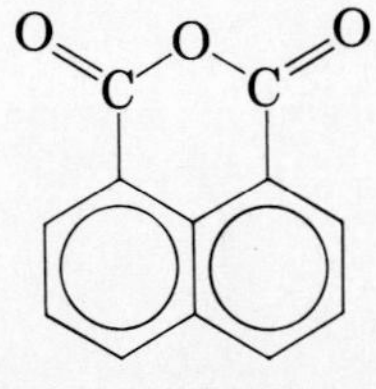

is related to the dicarboxylic acid named ______________________________

______________, and the anhydride itself is named ______________________________

______________________________.

Unsymmetrical acid anhydrides are named by using the name of each related acid (except for acid) as a separate word (alphabetically arranged) followed by the separate word anhydride.

9. The acid anhydride represented by the formula $C_6H_5—CO—O—CO—CH_3$ is related to two different carboxylic acids, commonly called ______________________________

and ______________________________. The name of the acid anhydride is

______________________________.

10. The unsymmetrical acid anhydride

$$CH_3—C_6H_4—SO_2—O—CO—CH_2—CH_2—CH_3$$

is named ______________________________.

ACID HALIDES

Compounds in which the OH group in an acid functional group is replaced by a halogen are called acid halides or acyl halides. They are named by two-word, functional class names. The group attached to halogen is named by modifying the name of the corresponding acid: for an acyclic acid name, the ending ic acid is changed to yl (for example, methanoic acid to methanoyl); for a cyclic acid name, the ending carboxylic acid is changed to carbonyl.

11. The acid halide $C_6H_5—CO—Cl$ is related to the acid named ______________________________

__________ and is itself named ______________________________.

12. Acetic acid has the formula ____________________, and acetyl chloride has the

formula ____________________.

13. The acid represented by the formula

$$CH_3—CH_2—\underset{\displaystyle Br}{\underset{|}{CH}}—COOH$$

is named ____, and the compound

$$\begin{array}{l} CH_3-CH_2-\underset{\displaystyle Br}{\underset{|}{CH}}-CO-Br \end{array}$$

is named ____.

14. 4-Nitrobenzenesulfonyl chloride has the formula

____________________________________.

15. The compound ▷—CO—Cl is named ____.

16. The compound

(naphthalene ring)—CO—Cl

is named ____ or
(trivial name)

____.
(systematic name)

ESTERS

Replacement of an OH group in an acid by an OR group (R = alkyl or aryl) gives an ester. If the ester is related to a carboxylic acid, the ester may be represented by the general formulas

$$R'-\overset{\displaystyle O}{\overset{\|}{C}}-O-R \quad \text{or} \quad R'-COOR \quad \text{or} \quad R-O-CO-R'$$

Note that the R group is always clearly attached to O and the R′ group to C=O in these illustrations. Esters may be related to acids other than carboxylic acids; for example, to nitric acid (acid, $HO-NO_2$; ester, $RO-NO_2$).

Esters are named in the same manner as salts (even though esters and salts are completely different from each other in properties): Two-word names are used. The R group (alkyl or aryl) is named as the first word, and the second word is formed by modifying the name of the acid to which the ester is related. The ending ic acid is replaced by ate. For example, acetic acid (ethanoic acid) and nitric acid will each form an ester, which will be called acetate (ethanoate) and nitrate, respectively.

17. Methanoic acid is more often called by its trivial name, ______________________.
The second word of the systematic name of an ester related to this acid will be ____________,
and the second word of a trivial name for the ester will be ______________.

18. Propionic acid is the trivial name for the acid whose systematic name is ____________
____________________. The second word of the trivial name of an ester related to this
acid will be ____________________, and of a systematic name, ____________________.

19. An ester formed from butyric acid will be called a ____________________.

20. Methyl butyrate has the formula ____.

21. The formula

$$\text{C}_5\text{H}_9-\overset{\overset{\displaystyle O}{\|}}{C}-O-\underset{}{\overset{\overset{\displaystyle CH_3}{|}}{C}H}-CH_2-CH_3$$

shows attached to oxygen an alkyl group named ____________________. The name of the alkyl group is the first word in the name of the ester. The acid to which the ester is
related is named ____, and the second word in the
name of the ester becomes ____. The complete
two-word name for the ester is ____.

22. A structural formula for trifluoroacetic acid is

______________,

a structural formula for the alkyl group isobutyl is

____,

and a structural formula for isobutyl trifluoroacetate is

____.

23. Isobutyric acid may be represented by the formula

______________________________,

and isobutyl isobutyrate by the formula

____.

24. A trivial name for

$$CH_2{=}CH{-}CH_2{-}O{-}\overset{\overset{\displaystyle O}{\|}}{C}{-}H$$

is______________________________.

25. Phenyl propionate may be represented by the structural formula

______________________________________,

and its isomer, benzyl acetate, by the formula

____.

26. To name a complex ester such as

$$CH_3{-}CH{=}\underset{\underset{\displaystyle Cl}{|}}{C}{-}CH_2{-}\underset{\underset{\displaystyle CH_3}{|}}{CH}{-}CH_2{-}COO{-}CH_2{-}\underset{\underset{\displaystyle O{-}CH_3}{|}}{CH}{-}CH_2{-}\underset{\underset{\displaystyle CH_3}{|}}{CH}{-}CH_3$$

we consider the alkyl group and the acyl group separately. The alkyl group of the ester is in the __________________________ (right or left) half of the formula shown, and the basis of its name will be a chain of ______ (number) carbons with substituents on positions numbered ____ and ____.

The substituents are named ____________ and ____________, respectively, and the name of the entire alkyl group is ______________________________.

The acyl group contains a parent chain of ______ (number) carbons with a chloro substituent on position number ____ and a methyl substituent on position number ____. The complete name of the carboxylic acid to which the ester is related is ______________________________, and the last word of the two-word ester name will be ______________________________. The complete name of the ester will be ______________________________.

27. The ester

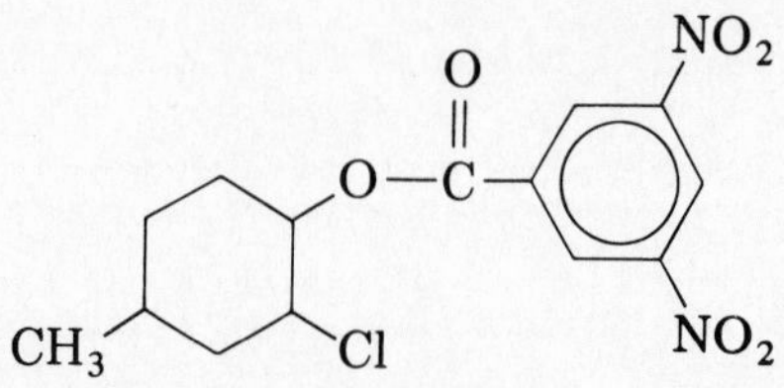

will have as the second word of its name ______________________________.

The alkyl group will be named ______________________________,

and the ester will be named ______________________________.

28. A structural formula representing <u>trans</u>-2-pentenyl <u>p</u>-toluenesulfonate is

______________________________.

29. The compound represented by the formula

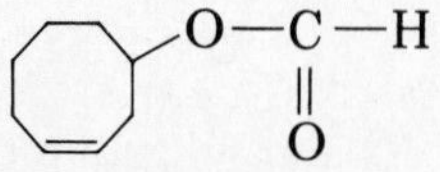

may be named ______________________________.

30. Boric acid may be represented by the formula $(HO)_3B$, and tributyl borate by the formula

____.

31. The ester

$$C_6H_5-SO_2-O-\bigcirc-\underset{CH_3}{\overset{CH_3}{\overset{|}{\underset{|}{C}}}}-CH_3$$

is related to an acid named ____. The second word of the two-word ester name is ____________________________________, and the full name of the ester is ____.

32. Vinyl trifluoromethanesulfonate has the formula

____________________________________.

33. A widely used detergent, sodium dodecyl sulfate, is both a salt and an ester related to the inorganic acid, ____________________acid. A structural formula for the compound (detergent) is

______________________________________.

34. The ester represented by the formula

$$CH_3-O-CH_2-CH_2-O-CH_2-CH_2-O-CO-CH_3$$

may be named readily with a replacement name. The replacement name of the alkyl group is ____; the name of the ester is

____.

When an ester group is to be named as a substituent (rather than serve as the basis of the name), it is named an alkyloxycarbonyl or aryloxycarbonyl group (sometimes shortened to alkoxycarbonyl or phenoxycarbonyl; see Chapter 7). For example, $CH_3—O—CO—$ as a substituent is called methoxycarbonyl. Note that a substituent with a combination name is always named toward the point of attachment, not from that point.

35. An acid group has higher priority than an ester group in naming (see Appendix). The name of the substituted acid

$$CH_3—CH_2—O—CO—C_6H_{10}—COOH$$

will include the substituent name, ______________________. The complete name of the acid illustrated is ____

______________________.

36. 4-Phenoxycarbonyl-1-naphthoic acid is represented by the formula

______________________________.

AMIDES

Compounds that contain an acyl group bonded to nitrogen are amides. Amides are named by replacing the ending in the name of the corresponding acid (ic acid, oic acid, or ylic acid, depending on the style of acid name used) by the systematic ending amide. For example, $CH_3—C(=O)—NH_2$ is related to acetic acid (ethanoic acid) and is named acetamide (ethanamide).

37. The compound represented by the formula $H—C(=O)—NH_2$ is related to the acid,

______________________ (trivial name), and is named ______________________.

38. Benzamide is represented by the formula

______________________,

and benzenesulfonamide by the formula

______________________________.

39. The formula

$$CH_3-C_6H_{10}-\overset{\overset{\displaystyle O}{\|}}{C}-NH_2$$

represents a compound related to the acid named ______________________________

______________________________. The amide is named ______________________________

______________________________.

Substituents may be on N as well as on positions in the acyl group. The locant N (underlined for italics) is used for substituents on N in exactly the same way as numerical locants are used for other substituents.

40. 4-Methylhexanamide is represented by the formula

______________________________,

and its isomer, N-methylhexanamide, is represented by the formula

______________________________.

41. N,N-Dimethylformamide is represented by the formula

______________.

42. The compound represented by the formula

$$C_6H_5-SO_2-NH-CH_2-CH_2-CH_2-CH_3$$

is named ______________________________.

43. The compound represented by the formula

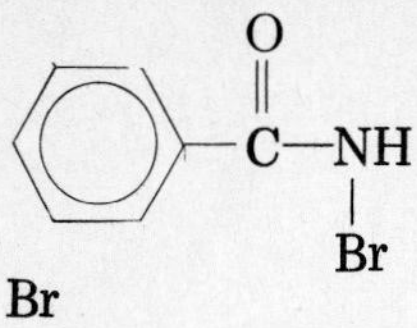

```
        O
        ‖
 ⌬—C—NH
          |
         Br
Br
```

has two substituents on a parent amide named ____________________. The positions of the substituents are designated by locants _____ and _____. The name of the substituted amide is ____________________.

11 Aldehydes and Ketones

The group $-\overset{\displaystyle O}{\overset{\|}{C}}-$ is called a carbonyl group. Compounds that contain a $-\overset{\displaystyle O}{\overset{\|}{C}}-H$ ($-CHO$) group attached to hydrogen or carbon are classified as aldehydes. Compounds that contain a carbonyl group attached to two carbons are classified as ketones. These two classes are related to each other in much the same way as primary and secondary alcohols are related to each other. Although a single systematic ending (ol) is used for the different classes of alcohols, chemists continue to use different systematic endings for aldehydes and ketones. Some nomenclature problems would be eased if a single ending were used for these carbonyl-containing compounds, and such practice for general use has been considered and is actually used for certain kinds of substituted compounds.

In aldehydes, at least one hydrogen is attached to the carbonyl group; in ketones, two hydrocarbon groups (that may be substituted) are attached to the carbonyl group.

1. The type of compound represented by the formula

$$CH_3-CH_2-\underset{\displaystyle O}{\underset{\|}{C}}-CH_3$$

is____________________.

2. The type of compound represented by the formula

$$CH_3-\overset{\displaystyle O}{\overset{\|}{C}}-H$$

is_________________________.

3. The formula

$$\begin{array}{c} H-C-H \\ \| \\ O \end{array}$$

represents ____________________, and
(type of compound)

O
‖
▷—C—◁

represents ____________________.
(type of compound)

For convenience in typing and writing, aldehydes are often represented by a condensed structural formula such as R—CHO. Note that the functional group is written —CHO rather than —COH, to avoid any confusion with alcohols. Ketones may be represented conveniently by a formula such as R—CO—R.

SYSTEMATIC (SUBSTITUTIVE) NAMES

Ketones and acyclic aldehydes are named by replacing the final e of the name of the corresponding hydrocarbon (including the carbon in the carbonyl group) with the systematic ending one (pronounced as in tone) or al (pronounced as in pal), respectively. For example, CH_3—CHO is ethanal, and CH_3—CO—CH_3 is propanone.

4. The formula CH_3—CH_2—CH_2—CH_2—CHO represents an aldehyde containing five carbons. The systematic name of the corresponding hydrocarbon is ______________, and the systematic name of the aldehyde is ____________.

5. Cyclohexanone may be represented by the formula

____________________.

6. Propanal may be represented by the formula

____________________.

Since the aldehyde functional group must necessarily be at the end of the chain, it will always be position number 1 in parent chains named as aldehydes. No number designation for the aldehyde functional group is necessary in the name. The carbonyl group in ketones, on the other hand, is not restricted to one position, and a locant is required in the name if isomeric positions are possible. When the carbonyl-containing compound is named as an alkanone, the ketone functional group is given the lower possible locant in the parent chain containing it. Substituent positions in both aldehydes and ketones are given the lower possible locants after the proper locant has been assigned to the carbonyl group.

7. There are two isomers which may be called pentanone; structural formulas for these

isomers may be written ______________________________ and

______________________________. The complete systematic names for these

isomers are ____________________ and ____________________, respectively.

8. A structural formula for 3-chloro-2-pentanone may be drawn

______________________________.

3-Chloropentanal may be represented by the formula

______________________________.

9. The compound whose formula is

$$\begin{array}{c} \quad\quad\quad\quad\quad\quad CH_3 \\ \quad\quad\quad\quad\quad\quad | \\ CH_3{-}CO{-}C{-}CH_3 \\ \quad\quad\quad\quad\quad\quad | \\ \quad\quad\quad\quad\quad\quad CH_3 \end{array}$$

is named ______________________________.

10. The hydrocarbon corresponding to the aldehyde

$$\begin{array}{l} CH_3{-}CH{-}CH_2{-}CH{=}CH{-}CHO \\ \quad\quad\quad\ | \\ \quad\quad\ \ C_6H_5 \end{array}$$

will have the substitutive name ________________________, and the systematic name for the aldehyde itself will be ________________________.

11. The hydrocarbon corresponding to the aldehyde

$$\begin{array}{l} CH_3-\underset{\displaystyle |\atop \displaystyle C_6H_5}{C}=CH-CH_2-CH_2-CH_2-CHO \end{array}$$

will actually have the substitutive name ________________________. When the compound represented by the formula is to be named as an aldehyde, however, the $-CHO$ functional group takes precedence over any other functional groups for numbering. The carbonyl group becomes position number ______, the alkene linkage is assigned position number ______, and the phenyl substituent appears on position number ______. The substitutive name for the aldehyde is ________________________.

12. Two alarm substances produced by ants are the compounds represented by formulas A and B.

$$CH_3-CH_2-CH_2-CH_2-CH_2-\underset{\displaystyle \| \atop \displaystyle O}{C}-CH_3$$

A

$$CH_3-\underset{\displaystyle |\atop \displaystyle CH_3}{C}=CH-CH_2-CH_2-\underset{\displaystyle |\atop \displaystyle CH_3}{CH}-CH_2-CHO$$

B

These naturally occurring compounds are named ________________________ (A)

and ________________________ (B).

13. The essence of caraway seed is a ketone represented by the formula

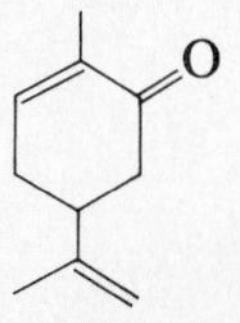

The parent compound on which the name of this ketone is based is

________________________, and the names of the substituents are

________________________ and ________________________.

The name of the ketone illustrated is ____
____.

14. Ethenone has the formula ______________________. It is more commonly called by the trivial name, ketene.

15. Even so complex an aldehyde as

CHO

which is involved in the chemistry of vision, can be named rather easily by a substitutive name. The cyclic substituent is on the last carbon of a parent chain of ________ (number) carbons; that substituent is named ______________________________________. Other substituents appear at positions __________ (numbers). The parent compound is named ______________
______________________________ (include specification of configuration), and the full name of the compound illustrated is ____
____________________. (Compare this name with the one in item 44, Chapter 6.) The common name for this aldehyde is all-trans-retinal.

16. The compound

$$-CH_2-\underset{\underset{O}{\|}}{C}-CH_2-CH_2-CH_3$$

is named as a substituted alkanone. The substituent, on position number ________, is named ____________________. The name of the ketone illustrated is
____.

The name of the isomer of the foregoing ketone,

$$-\underset{\underset{O}{\|}}{C}-CH_2-CH_2-CH_2-CH_3,$$

is 1-cyclohexyl-1-pentanone, even though the parent compound, 1-pentanone, is actually an aldehyde and is named pentanal. The one ending is the approved one for all similar compounds and clearly conveys the correct functional group information. This use illustrates the appeal of a single systematic ending for aldehydes and ketones.

17. 1,4-Diphenyl-1-hexanone is represented by the formula

____.

18. The ketone

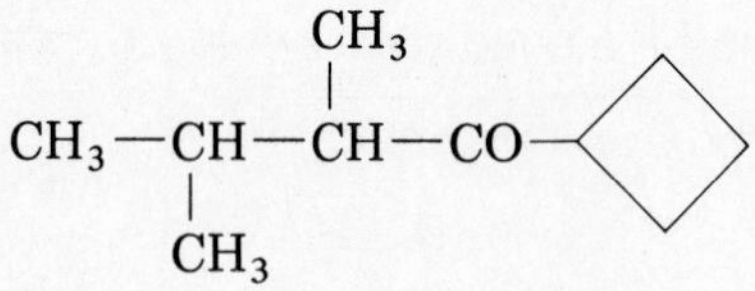

```
                 CH3
                  |
CH3—CH—CH—CO—◇
         |
        CH3
```

is named ____.

Compounds containing two like carbonyl groups are named by combining the ending dial (two syllables) or dione with the name of the corresponding parent hydrocarbon. Locants are used for the carbonyl groups in names of diones but are not used in names of dials (the two CHO groups are the termini of the parent chain).

19. Hexanedial is represented by the formula ______________________, and 2,4-hexanedione by the formula ______________________.

20. C_6H_5—$CH(CH_2$—CH_2—$CHO)_2$ is a condensed formula representation of the aldehyde named ____.

CYCLIC ALDEHYDES

When one or more —CHO groups are attached directly to a cyclic system, the compound is named by adding the ending carbaldehyde, dicarbaldehyde, etc., to the name of the cyclic system. Except for cyclic systems with fixed numbering (naphthalene, for example), the carbon to which a —CHO group is attached is the number 1 position, and a locant for the —CHO group in a monoaldehyde is not needed or used. For example, ▷—CHO is cyclopropanecarbaldehyde. Locants for all —CHO groups in a polyaldehyde are required; they precede the name of the cyclic system, as usual.

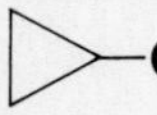

Chemical Abstracts uses the ending carboxaldehyde rather than carbaldehyde, but the IUPAC ending avoids the redundancy of both ox and aldehyde indicating oxygen.

21. The name for

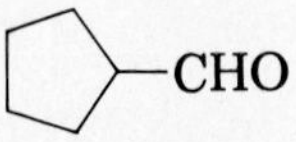

(cyclopentane ring)—CHO

is ______________________________, and the name for its isomer,

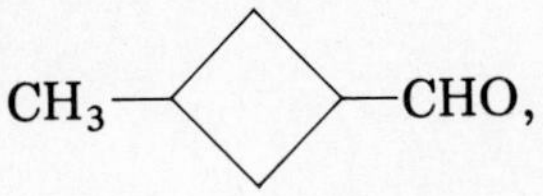

CH_3—(cyclobutane ring)—CHO,

is ______________________________.

22. 2,6-Naphthalenedicarbaldehyde is represented by the formula

______________________________.

23. The systematic name for

O_2N—(benzene ring)—CHO

is ______________________________.

TRIVIAL NAMES OF ALDEHYDES

Several simple aldehydes are commonly named by IUPAC-approved names based on the approved trivial names of the corresponding carboxylic acids (see Chapter 9). The ending aldehyde replaces ic acid or oic acid in the trivial name of the acid. For example, H—COOH is formic acid, and H—CHO is formaldehyde.

24. Acetic acid has the formula ____________________, and acetaldehyde has the formula ____________________.

25. The trivial name for the acid corresponding to the aldehyde

$$\begin{array}{c} CH_3 \\ | \\ CH_3—CH—CHO \end{array}$$

is ______________________________, and the trivial name for the aldehyde is

______________________________.

26. Benzenecarbaldehyde is the systematic name for the aldehyde whose formula is

____________________. The trivial name of the acid corresponding to this aldehyde

is ______________________________, and the trivial name of the aldehyde itself is

______________________________.

27. The trivial name of the aldehyde CCl_3—CHO is ______________________________.

FUNCTIONAL CLASS NAMES OF KETONES

Some ketones are usually named by multiple-word names ending in ketone. The ketone portion of the name refers to the carbonyl group, and the two groups attached to the carbonyl group are named alphabetically and separately. If both groups are alike (symmetrical ketones), the multiplying prefix di is used with the name of the group.

28. The formula for ethyl methyl ketone is ______________________________. The preferred name for this ketone is the systematic name, 2-butanone.

29. The functional class name for $\triangleright$—CO—$\triangleleft$ is ______________________________

____________________, and that for C_6H_5—CO—C_6H_5 is ______________________________

____________________. Symmetrical ketones of this type (two cyclic groups) are named by *Chemical Abstracts* with systematic names based on methanone as the parent compound, that is, diphenylmethanone instead of diphenyl ketone. Note again that the unsubstituted parent compound in this example is actually an aldehyde.

30. The ketone

—CO—CH_2—CH_3

may be named with the functional class name, ______________________________

__________, or with the systematic name, ______________________________.

31. Benzyl sec-butyl ketone has the formula

____

and the systematic name ______________________________.

TRIVIAL NAMES OF KETONES

A few ketones are commonly known by names containing the stems of the trivial names of related carboxylic acids. Acetone (CH_3—CO—CH_3) is an example of this kind of name and is the only trivial name for an acyclic ketone whose continued use is not discouraged. Trivial names continue to be used for some unsymmetrical ketones in which one of the groups is phenyl or naphthyl. For these ketones, the endings ophenone or onaphthone, respectively, replace the ic acid or oic acid ending of the trivial name of the carboxylic acid corresponding to the remainder of the ketone (that is, the acyl group). (Locants in the aryl group are primed to distinguish them from the unprimed locants in the acyl group.) For example, CH_3—CO—C_6H_5 is commonly known as acetophenone. The shortened stem propi rather than propion for CH_3—CH_2—CO— is used in this kind of name.

32. 2′-Propionaphthone is represented by the formula

________________________.

Two other names for this compound are used in item 30.

33. The trivial name for C_6H_5—CO—C_6H_5 is ________________________.

34. 2,2,2-Trifluoroacetophenone has the formula

________________________.

CARBONYL GROUPS AS SUBSTITUENTS

Some aldehydes and ketones contain other functional groups which may be chosen as the basis of the name of the compound. For example, a compound may contain both a ketone functional group and a carboxylic acid functional group. (The relative priorities of functional groups for naming multifunctional compounds are given in a table in the Appendix; in general, the more bonds from carbon to hetero atoms in the functional group, the higher the priority of that functional group for naming.) If another functional group has priority as the basis of the name, the oxygen of the aldehyde or ketone carbonyl group is treated as a substituent. The prefix oxo, signifying the =O substituent (not the carbon also), is used just as any other substituent prefix; it does not have any special preference in assignment of locants. Note that the prefix for =O as a substituent on a parent chain is oxo, while that for —O— as a replacement for carbon in a parent chain is oxa.

35. 4-Oxocyclohexanecarboxylic acid has the formula

______________________________, and 4-oxacyclohexanecarboxylic acid has the formula

______________________________.

36. The structural formula for 6-oxohexanoic acid may be drawn

____.

In addition to being a carboxylic acid, this compound may also be classified as a(n)

______________________.

37. "Queen substance," a bee sex attractant isolated from queen honey bee glands, is 9-oxo-*trans*-2-decenoic acid, which may be represented by the structural formula

____.

38. The compound represented by the formula

may be named as an aldehyde; the basis of the name, without specifying numbering or substituent, will be ______________________________, and proper numbering of the ring will assign position number ____ to the alkene linkage. The complete name for the substituted aldehyde will be ____.

A —CHO group to be treated as a substituent is named formyl or methanoyl.

39. The compound represented by the formula

CHO

$COOCH{=}CH_2$

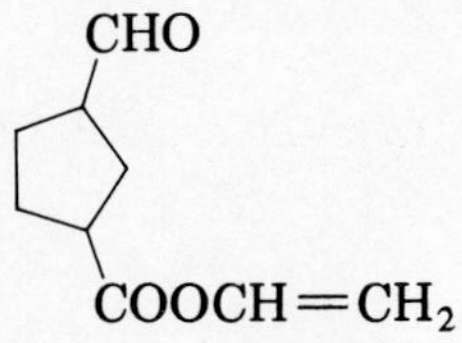

is named as a substituted ester. It is named

____.

40. 4-Formylbenzoic acid may be represented by the formula

______________________________.

12 Amines and Related Cations

Amines are organic compounds related to ammonia, both in structure and chemical behavior. In fact, in organic nomenclature, amine is synonymous with and is used in place of ammonia. The trivalent nitrogen atom in amines is bound only to hydrocarbon groups (that may be substituted) and hydrogens. If only one hydrogen of ammonia (amine) is replaced by a hydrocarbon group (for example, $CH_3—NH_2$), the amine is classified as a primary amine; if two hydrocarbon groups are bound to the nitrogen, the amine is classified as a secondary amine; and if three hydrocarbon groups are bound to nitrogen (no hydrogens bound to nitrogen), the amine is a tertiary amine.

1. The compound represented by the formula $CH_3—CH_2—NH—CH_2—CH_3$ is

classified as a ______________ amine.

2. The compound represented by the formula

$$\begin{array}{c} CH_3 \\ | \\ CH_3—N—CH_3 \end{array}$$

is classified as a ______________ amine.

3. When hydrocarbon groups are represented by the general symbol, R, primary amines may be represented by the generalized formula ______________, secondary amines by the generalized formula ______________, and tertiary amines by the generalized formula

______________.

Note that the classification of amines depends on the degree of substitution on the nitrogen atom, not on the nature of the hydrocarbon group or groups. The classification of alcohols, on the other hand, depends on the nature of the hydrocarbon group, not on the oxygen, since only one group can be attached to oxygen in an alcohol.

4. The alcohol represented by the formula

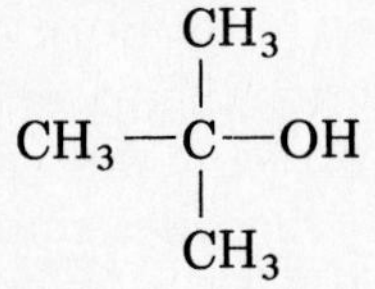

$$\begin{array}{c} \quad CH_3 \\ \quad | \\ CH_3-C-OH \\ \quad | \\ \quad CH_3 \end{array}$$

is classified as a ______________ alcohol, and the amine represented by the formula

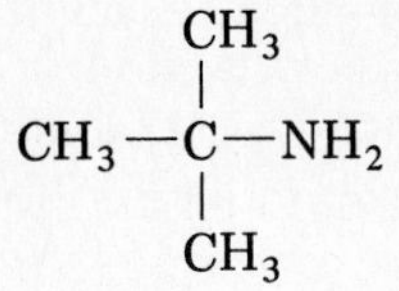

$$\begin{array}{c} \quad CH_3 \\ \quad | \\ CH_3-C-NH_2 \\ \quad | \\ \quad CH_3 \end{array}$$

is classified as a ______________ amine.

5. The alcohol represented by the formula

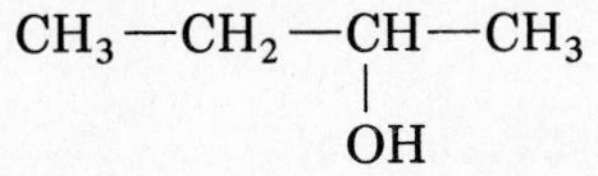

$$\begin{array}{l} CH_3-CH_2-CH-CH_3 \\ \qquad\qquad\quad | \\ \qquad\qquad\ OH \end{array}$$

is classified as a ______________ alcohol, and the amine represented by the formula

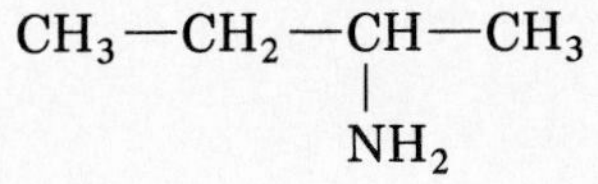

$$\begin{array}{l} CH_3-CH_2-CH-CH_3 \\ \qquad\qquad\quad | \\ \qquad\qquad\ NH_2 \end{array}$$

is classified as a ______________ amine.

6. The alcohol represented by the formula

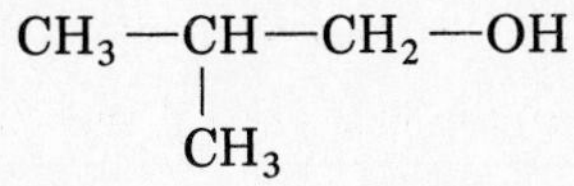

$$\begin{array}{l} CH_3-CH-CH_2-OH \\ \qquad\quad | \\ \qquad\ CH_3 \end{array}$$

is classified as a ______________ alcohol, and the amine represented by the formula

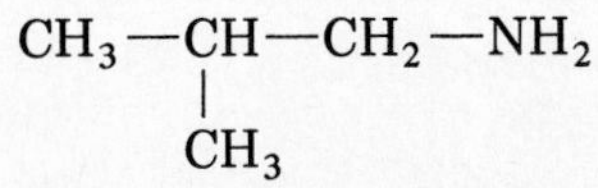

$$\begin{array}{l} CH_3-CH-CH_2-NH_2 \\ \qquad\quad | \\ \qquad\ CH_3 \end{array}$$

is classified as a ______________ amine.

Primary amines are named by one-word substitutive names formed in either of two ways: (1) On the basis of the parent hydrocarbon to which $—NH_2$ is attached. The final e of the name of the parent hydrocarbon is replaced by the systematic ending amine; a locant included in the usual way specifies the position of attachment of $—NH_2$. For example, $CH_3—CH_2—CH_2—NH_2$ is 1-propanamine. *Chemical Abstracts* uses this style. These names parallel the substitutive names of alcohols (for example, 1-propanol); the amine functional group takes precedence over substituents for numbering the parent chain. (2) On the basis of the parent compound, amine (synonymous in this context with ammonia, NH_3). The name of the hydrocarbon group attached to $—NH_2$ modifies amine in a one-word name. Amine is the name of an individual compound, NH_3. No locant is needed or used, because the hydrocarbon group can be attached only to N. For example, $CH_3—CH_2—CH_2—NH_2$ is propylamine. *Chemical Abstracts* does not use this style.

The first style (based on parent hydrocarbon) is recommended for all primary amines, but the second style is more generally used for amines of simple structure.

Symmetrical secondary and tertiary amines with simple groups (all alike) attached to N are named by style (2) above with the inclusion of the appropriate multiplying prefix (di, tri). For example, $(CH_3)_3N$ is trimethylamine.

7. The amine represented by the formula

$$\begin{array}{l} CH_3—CH_2—CH—NH_2 \\ \qquad\qquad\quad\;\; | \\ \qquad\qquad\quad CH_3 \end{array}$$

may be named by two names: one based on parent hydrocarbon,

______________________________, and one based on parent amine,

______________________________.

8. tert-Butylamine may be represented by the formula

____________________.

9. $C_6H_5—NH—C_6H_5$ may be named ______________________________.

10. (1-Ethylpentyl)amine, represented by the formula

______________________________,

may also be named ______________________________. The latter name is preferred.

11. The compound

$$CH_3-\underset{}{\overset{CH_3}{\overset{|}{C}H}}-CH_2-\underset{\underset{CH_3-CH_2}{|}}{\overset{\overset{CH_3}{|}}{C}}-CH_2-CH_2-NH_2$$

is preferably named on the basis of the parent hydrocarbon. The name is

____.

12. The compound

$$(CH_3)_2CH-C_6H_{9}(NH_2)(C_6H_5)$$

(cyclohexane ring bearing NH_2 and C_6H_5 on adjacent carbons, and $(CH_3)_2CH$ on another ring carbon)

may be named on the basis of a parent hydrocarbon. The name is

____.

13. Tricyclopropylamine may be represented by the formula

______________.

Unsymmetrical secondary and tertiary amines (all hydrocarbon groups not alike) are named as N-substituted derivatives of a primary amine. The primary amine with the largest or most complex hydrocarbon group is chosen as the parent amine for the name, and the other groups on N are treated as substituents with locant N (capital; underlined for italics). For example, $CH_3-CH_2-NH-CH_3$ is named N-methylethanamine. IUPAC rules permit the name N-methylethylamine, but the parent hydrocarbon-type name is preferred.

14. The amine represented by the formula

$$CH_3-\underset{\underset{CH_3}{|}}{CH}-CH_2-NH-\underset{\underset{CH_3}{|}}{CH}-CH_3$$

may be named as a derivative of the primary amine, ______________________.

The unsymmetrical amine may also be named ______________________.

15. The parent hydrocarbon on which a systematic name of $C_6H_5—N(CH_3)_2$ will be based is named ______________________, and the tertiary amine $C_6H_5—N(CH_3)_2$ may be named ______________________________. This compound is more frequently identified by its IUPAC-approved trivial name, N,N-dimethylaniline, but *Chemical Abstracts* uses the systematic, parent hydrocarbon name in indexes.

16. N,N-Dimethyl-4-vinylcyclohexylamine may be represented by the formula

______________________________________. The name used here is based on ______________________ as parent compound.

17. N-Phenyl-4-nitro-1-naphthalenamine is represented by the formula

______________________________.

Note that the name 3-pentanamine is correct (preferred), because the parent, pentane, has a position 3 to which the functional group, amine, can be attached, whereas the name 3-pentylamine is incorrect, because 3-pentyl is not a correct name for the alkyl substituent on the parent, amine.

The amine functional group is low in the priority order for citation as a suffix (see the Appendix for order of precedence), but when amine is the ending of an IUPAC name, the $—NH_2$ group (amino group) takes precedence over other substituents or functional groups for numbering the parent chain.

18. The preferred name (based on parent hydrocarbon) for

$$\begin{array}{cccccc} CH_3— & CH— & CH_2— & CH_2— & CH— & CH_3 \\ & | & & & | & \\ & CH_3 & & & NH_2 & \end{array}$$

will indicate the amine functional group on carbon number ______. A complete name for the amine will be ____.

19. The parent hydrocarbon name (without locant) on which the name of the amine

$$\begin{array}{c} \quad\quad\quad\quad O-C_6H_5 \\ \quad\quad\quad | \\ CH_3-CH_2-CH-CH=CH-CH_2-CH_2-CH_2-N(CH_3)_2 \end{array}$$

is based is ______________________, and the name of the amine itself is ______________________.

20. A name, based on hydrocarbon parent, for the amine represented by the formula

CH_3-O-(cyclooctene ring)$-NH_2$

is ______________________.

21. The formula

$$\begin{array}{l} CH_3-CH_2-CH-CH=C-CH_3 \\ \quad\quad\quad\quad\quad\ \ | \quad\quad\quad\quad | \\ \quad\quad\quad\quad\ \ CH_3 \quad\quad\ N-CH_2-CH_3 \\ \quad\quad\quad\quad\quad\quad\quad\quad\quad\ \ | \\ \quad\quad\quad\quad\quad\quad\quad\quad\ CH_3 \end{array}$$

indicates a substituted amine which is preferably named on the parent hydrocarbon basis. Locants for the methyl substituents are ______________, and the locant for the ethyl substituent is ______. The complete name for this substituted amine is

______________________.

In the names of compounds that contain other functional groups with higher priority as the basis of the name, the $-NH_2$ functional group is treated as a substituent named amino. In the IUPAC order of priority for name suffixes, amines are lower than nearly all other functional groups.

22. The compound represented by the formula

$$\begin{array}{l} CH_3-CH_2-CH-CH-CH_2-CH_3 \\ \quad\quad\quad\quad\quad\ \ | \quad\quad | \\ \quad\quad\quad\quad\ \ OH \quad NH_2 \end{array}$$

is named as an alcohol. The substituent will be called ______________, and the full name of the compound will be ______________________.

23. The compound represented by the formula

$$(CH_3)_2N-CH_2-\underset{\displaystyle CH_3}{\underset{|}{CH}}-CH_2-CO-CH_3$$

is named as a ketone with a methyl substituent on position number ______ and a ______________________________ substituent on position number ______. The complete substitutive name for this ketone is ____.

24. A structural formula for ethyl 2-diethylamino-3-ethoxydecanoate may be drawn

____.

Compounds containing more than one amine functional group (amino group) attached to a parent hydrocarbon are named by use of the appropriate multiplying prefix between the ending amine and the name of the parent hydrocarbon. (Note that the final e of the hydrocarbon name is retained as usual before a consonant.) These names parallel those for alkanediols. For example, $H_2N-CH_2-CH_2-CH_2-NH_2$ is 1,3-propanediamine.

25. Products produced by decaying flesh are $H_2N-CH_2-CH_2-CH_2-CH_2-NH_2$ and $H_2N-CH_2-CH_2-CH_2-CH_2-CH_2-NH_2$, sometimes called putrescine and cadaverine, respectively. IUPAC names (more systematic and prosaic) for these amines are ______________________ and ______________________, respectively.

26. 1,6-Hexanediamine, used in the manufacture of nylon, may be represented by the formula ____.

27. 3-Methyl-2,4-pentanediamine may be represented by the formula

______________________________.

28. The triamine

$$H_2N-CH_2-\underset{\displaystyle NH_2}{\underset{|}{CH}}-CH_2-CH_2-CH_2-NH_2$$

is named ____.

29. N,N,N′,N′-Tetramethyl-1,2-ethanediamine has the formula

______________________________________. 1,2-Ethanediamine, whose formula is

______________________________________, is often called by the bivalent name, ethylenediamine.

AMINIUM AND AMMONIUM COMPOUNDS

When ammonia is protonated ($NH_3 \rightarrow \overset{+}{N}H_4$), an ion called ammonium ion is generated. The ions formed by protonation of amines have also been called ammonium ions, but the designation aminium ions is now recommended. The suffix ium is coming to be accepted to mean, in general application, "plus a proton." Thus aminium means "amine plus a proton." Extensions of this meaning for ium will be found in Chapter 14. Aminium compounds are named with multiple-word names; the cation and the anion are named separately. For example, $CH_3-\overset{+}{N}H_3\ \ \overset{-}{Cl}$ is methanaminium chloride (or methylaminium chloride).

30. The amine represented by the formula $(CH_3-CH_2)_2NH$ is usually named ____________

______________, and the salt derived from it, $(CH_3-CH_2)_2\overset{+}{N}H_2\ \ \overset{-}{Cl}$, is named

____.

31. 2-Naphthalenaminium perchlorate is represented by the formula

______________________________________.

32. 4-tert-Butyl-N,N-dipropylcyclohexanamine is converted by protonation into the ion

named ____.

33. Hydroxylamine, $HO-NH_2$, a commonly used reagent for aldehydes and ketones, is usually marketed as the salt under the name hydroxylamine hydrochloride. This salt can

be represented by the formula ______________________ and more systematically named ______________________.

The neutral compounds commonly classified as amino acids are internal salts (zwitterions) which are properly named by citing the cationic substituent aminio (o replaces um) as a prefix and the anionic component as a suffix (one-word names). For example, $H_3\overset{+}{N}-CH_2-CO\overset{-}{O}$ is aminioacetate (or aminioethanoate). Note that amino is the name of an unchanged $-NH_2$ substituent, and aminio is the name of a cationic $-\overset{+}{N}H_3$ substituent.

34. The amino acid commonly known as alanine is systematically named 2-aminiopropionate. It has the formula

______________________.

35. The amino acid leucine is 2-aminio-4-methylpentanoate and has the formula

______________________.

36. Threonine has the formula

$$CH_3-\underset{\displaystyle OH}{\underset{|}{CH}}-\underset{\displaystyle \underset{+}{NH_3}}{\underset{|}{CH}}-COO^-$$

and the systematic name ______________________.

Cations that contain a nitrogen bound to four hydrocarbon groups are called quaternary ammonium ions. Aminium ions have at least one hydrogen bound to nitrogen; quaternary ammonium ions have none. Quaternary ammonium ions and compounds are named by preceding ammonium by the names of the four groups covalently bound to N, alphabetically arranged, all in one word; the name of the anion follows as a separate word.

37. Benzyltrimethylammonium chloride can be represented by the formula

______________________.

38. The amine $[CH_3-(CH_2)_6-CH_2]_2N-CH_2-CH_3$ is named

____,

and the salt, $[CH_3-(CH_2)_6-CH_2]_2\overset{+}{N}(CH_2-CH_3)_2\ \bar{I}$, is named

____.

39. With R symbolizing an alkyl group, a quaternary ammonium hydroxide may be

represented by the generalized formula ____________________.

40. The biologically important quaternary ammonium hydroxide, $HO-CH_2-CH_2-\overset{+}{N}(CH_3)_3\ \overset{-}{O}H$, is commonly called <u>choline</u>. Its systematic name is

____.

Some biologically important compounds containing a quaternary ammonium group are zwitterions. These internal salts are properly named by citing the cationic substituent <u>ammonio</u> (<u>o</u> replaces <u>um</u>) as a prefix and the anionic component as suffix (one-word names). The hydrocarbon groups bound to the ammonio nitrogen must be included in the name of the substituent as usual.

41. The zwitterion commonly known as betaine (three syllables) has the formula $(CH_3)_3\overset{+}{N}-CH_2-CO\overset{-}{O}$ and the systematic name

____.

13 Bridged Ring Systems

Hydrocarbon systems that have two or more carbons common to two or more rings are called bridged hydrocarbons. A combination of a multiplying prefix (bi for two) and the prefix cyclo specifies the number of rings. For example, a bicycloalkane is a saturated hydrocarbon whose structure is two rings joined through two common carbon atoms. The carbon atoms at the junctures of the rings are called bridgeheads, and the bonds, atoms, or chains of atoms connecting the bridgeheads are called bridges.

```
          H
          |
          C  <------- bridgehead (one of two in illustration)
        / | \
    H2C   |   CH2  <-- bridge (one of three in illustration)
        \ | /
          C
          |
          H
```

Although bridged ring systems containing more than two rings are well known, this chapter will be confined to bicyclo compounds.

The stem that replaces alk in the specific name of the bicycloalkane corresponds to the total number of carbons in the two rings. The formula above represents a bicyclobutane.

1. The hydrocarbon represented by the formula

```
CH2—CH—CH2
|    |     \
|   CH2     CH2
|    |     /
CH2—CH—CH2
```

is a bicycloalkane containing a total of ________ (number) carbons in the two rings; it will be called a ______________________.

2. The hydrocarbon represented by the formula

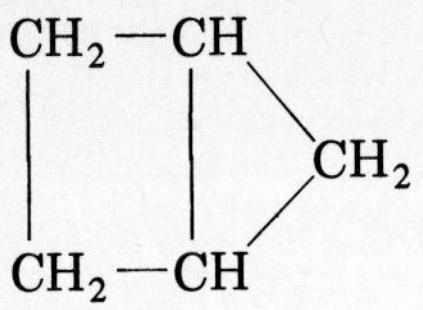

$$\begin{array}{l} CH_2-CH \\ \;|\qquad\;\, |\;\;\diagdown \\ \;|\qquad\;\, |\qquad CH_2 \\ \;|\qquad\;\, |\;\;\diagup \\ CH_2-CH \end{array}$$

will be called a ________________________.

3. The two bridgeheads in the bicyclooctane formula of item 1 are separated from each other by three bridges, containing ____, ________, and ____ carbons, respectively.
(numbers)

4. The two bridgeheads in the bicyclopentane formula of item 2 are separated from each other by three bridges, containing ____, ________, and ____ carbons, respectively.
(numbers)

These numbers, designating the length of the bridges, are used in the full name of a bicycloalkane to differentiate it from isomeric bicycloalkanes. The style is illustrated by the name bicyclo[3.2.1]octane. Note that the numbers are arranged in descending order, are separated from each other by periods, and are enclosed in brackets. The name is a one-word name without any space separation between parts.

5. The full name for the bicyclopentane illustrated in item 2 is

________________________________.

Bicycloalkanes, like cycloalkanes, are conveniently represented by geometric figures. The formulas

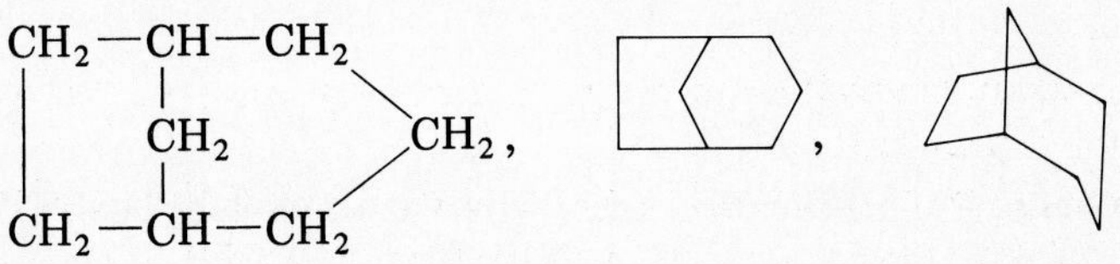

$$\begin{array}{l} CH_2-CH-CH_2 \\ \;|\qquad\;\;\, |\qquad\;\;\diagdown \\ \;|\qquad\; CH_2\qquad\quad CH_2, \\ \;|\qquad\;\;\, |\qquad\;\;\diagup \\ CH_2-CH-CH_2 \end{array}$$

are equivalent representations of bicyclo[3.2.1]octane. The last figure is intended to show the actual geometry of the molecule.

Several isomeric bicyclooctanes may be represented by formulas, five of which are

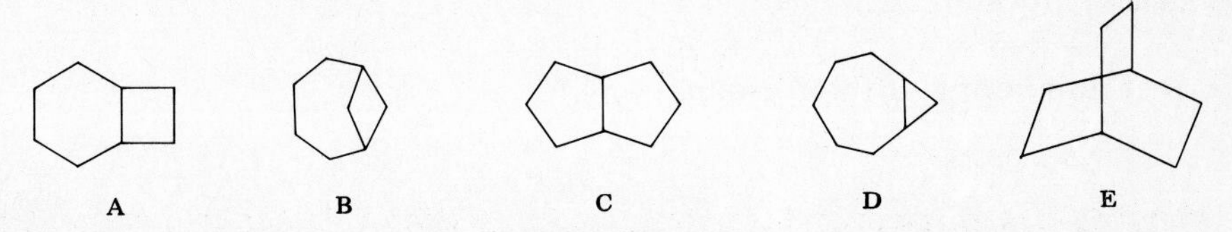

These isomeric bicyclooctanes are differentiated from one another by full names that include numbers.

6. The distinguishing, full name for

formula A is ______________________________;

formula B is ______________________________;

formula C is ______________________________;

formula D is ______________________________;

formula E is ______________________________.

Note that the numbers in brackets account for all carbons other than the bridgeheads in the bicycloalkane framework; the sum of these numbers always equals two less than the number of carbons signified by the stem in the name.

Numbering of a bicycloalkane to indicate location of substituents begins at one bridgehead, proceeds around the longest bridge to the other bridgehead, continues around the second longest bridge back to the number 1 position (original bridgehead), and is completed across the shortest bridge. The numbered formula for bicyclo[3.2.1]octane illustrates the numbering.

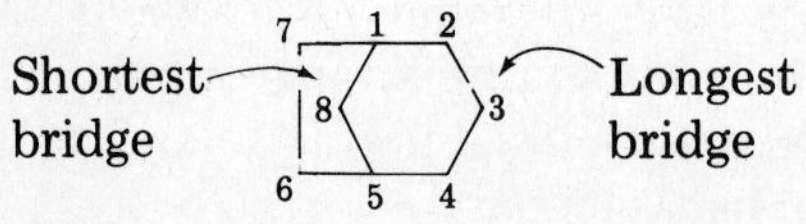

Shortest bridge 7 1 2 8 3 6 5 4 Longest bridge

The choice of bridgehead for position number 1 is made to permit substituents to be assigned the smaller possible position numbers.

7. Correct numbering of the formula

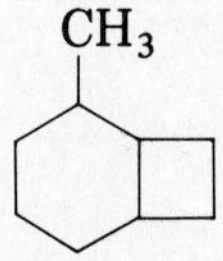

CH_3

will assign to the methyl substituent locant _____, whereas the methyl substituent in

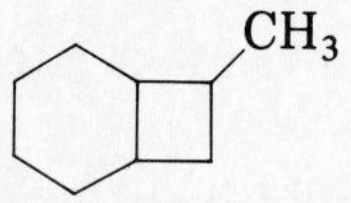

CH_3

will be assigned locant _____.

8. Correct numbering of the formula

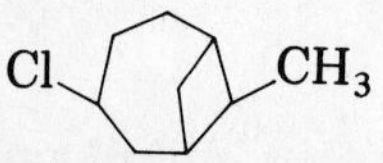

Cl CH_3

will assign to the methyl substituent locant _____ and to the chloro substituent locant _____.

9. 8,8-Dichlorobicyclo[5.1.0]octane may be represented by the formula

______________________________.

10. 1,3-Dimethylbicyclo[1.1.0]butane may be represented by the formula

______________________________.

As in naphthalene, the numbering of bicycloalkanes is fixed, and the locants for functional groups follow from that fixed numbering. Bridged hydrocarbons containing carbon-carbon double bonds in the ring system are named by replacing the final ending ane with ene and inserting a locant immediately before ene to indicate position of the alkene linkage. The bridgehead that will permit the alkene linkage to have the smaller locant is chosen for position number 1.

11. The alkene represented by the formula

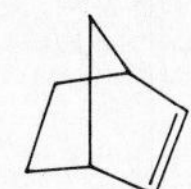

may be named with the systematic name ______________________________.

12. α-Pinene (trivial name), the major constituent of turpentine, is 2,6,6-trimethylbicyclo[3.1.1]hept-2-ene; α-pinene may be represented by

______________________________.

13. Carene, an isomer of α-pinene, may be represented by the formula

CH_3 CH_3

CH_3

The systematic name for carene is

______________________________.

Bridged ring systems containing other functional groups may be named in much the same way as are bicycloalkanes; the ending ane is replaced with the appropriate systematic ending, which is preceded by the smaller possible locant for that functional group.

14. The alcohol

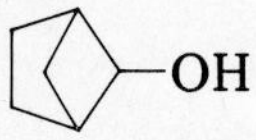

may be named by the systematic name ______________________________.

15. Camphor is a naturally occurring ketone represented by the formula

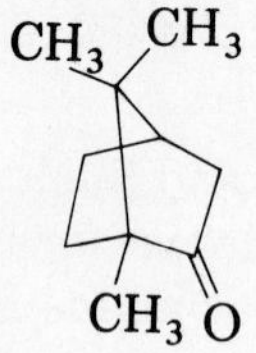

A systematic name for camphor is

____.

16. Bicyclo[4.4.0]decane-2-carboxylic acid may be represented by the formula

______________________.

HETEROCYCLIC BRIDGED SYSTEMS

Some bridged ring systems containing ring atoms other than carbon are named conveniently by replacement names which include a locant and a prefix (oxa for O, thia for S, aza for N) to identify each replacing hetero atom.

17. 7-Oxabicyclo[4.1.0]heptane is the name by which *Chemical Abstracts* indexes the compound commonly called cyclohexene oxide and represented by the formula

______________________.

18. The compound represented by the formula

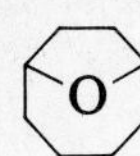

may be named with the systematic name ______________________________.

19. The name for the hydrocarbon represented by the formula

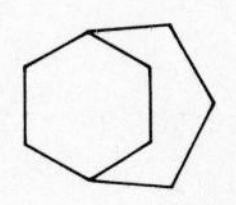

is ______________________________; the prefix signifying replacement of carbon by nitrogen is ________; and the systematic name for the compound represented by the formula

is ______________________________.

20. 1-Azabicyclo[2.2.2]octane (sometimes called quinuclidine) may be represented by the formula

______________________________,

and 7-thiabicyclo[2.2.1]heptane by the formula

______________________________.

21. Cocaine, a local anesthetic that is now considered addictive, has the structure

CH_3

N

$COOCH_3$

$OCO—C_6H_5$

and is systematically named as an ester. The substituent $-O-CO-C_6H_5$ is named benzoyloxy and is on position number ______. The parent acid (without substituents) is named ______________________________, and the ester illustrated is named

______________________________.

BICYCLO[2.2.1] HEPTANE (NORBORNANE) DERIVATIVES

Derivatives of bicyclo[2.2.1]heptane has been cited so frequently in the chemical literature that some special attention to nomenclature of these derivatives is appropriate. The IUPAC-approved trivial name for 1,7,7-trimethylbicyclo[2.2.1]heptane is bornane. In polycyclic compounds of this class, "nor" is a prefix used to indicate that all carbons outside the ring system of the parent compound are missing; only the ring skeleton of the parent compound remains.

22. Bornane is the trivial name for the compound represented by the formula

______________________,

and norbornane is the trivial name for the compound represented by the formula

______________________.

Norbornane is so widely used as a name for bicyclo[2.2.1]heptane that most chemists probably do not even make a mental reference back to the parent compound, bornane, when the name norbornane appears. Numbering of positions in norbornane follows the rules for all bicycloalkanes, and substituted norbornanes are named just as other substituted hydrocarbons are.

23. The nitro substituent in the formula

NO_2

is located on position number ______, and the name for the compound represented by the formula is ______________________.

24. 7-Chloronorbornane may be represented by the formula

______________________________.

25. 2-Norbornene is represented by the formula

______________________________,
and the unsaturated compound represented by the formula

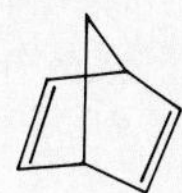

is named ____.

26. 2-Norbornanol is represented by the formula

______________________________, and

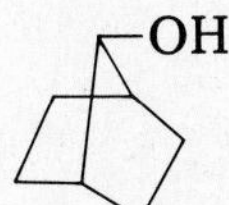

is named __________________________.

27. Camphor has the formula

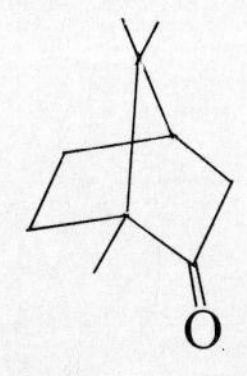

and norcamphor has the formula

______________.

Norcamphor can also be named as a derivative of norbornane, that is,

_________________________.

28. 1-Methylnorbornane may be represented by the formula

_________________________.

The geometry of some substituted bicycloalkanes (such as norbornanes) is such that a substituent on the main ring may extend "within" or "outside" the obtuse angle of the main ring. The italicized prefixes *endo* ("within") and *exo* ("outside") are used to differentiate such isomers.

29. *exo*-2-Norbornanol may be represented by the formula

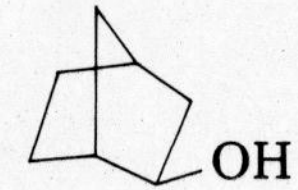

and *endo*-2-norbornanol by the formula

_________________________.

30. Br is the formula for ____.

31. 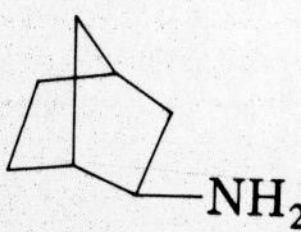

may be named ____.

32. exo-2-Norbornyl acetate may be represented by the formula

____,

and endo-2-norbornyl benzenesulfonate by the formula

____.

14 Nomenclature of Reaction Intermediates

The transient nature of reaction intermediates does not reduce the need for precise nomenclature for them. Yet the rules of nomenclature for these neutral and charged species are less systematized than they are for the more stable classes of compounds treated earlier in this book. Because these rules are occasionally inconsistent with other rules and practices, some, in use, actually imply the incorrect structure. But revision in the rules of chemical nomenclature is slow at best. Committees of IUPAC and the American Chemical Society have concerned themselves with the nomenclature of reaction intermediates, and proposed revisions in the rules are being considered. These recommendations have guided the emphases in this chapter. The styles of names set forth are those which are gaining favor because of their simplicity and consistency. They may in time become (when they are not already) the ones advocated by the authority of IUPAC. Meanwhile, they can be used with the confidence that they facilitate rather than impair accurate communication of structural information.

FREE RADICALS

Neutral organic species that contain at least one unpaired electron are called free radicals. Structural formulas for free radicals generally include a dot for the unpaired electron close to the symbol for the atom with that electron. Free radicals are named by using the name of the group as though it were a substituent, together with the separate word "radical"; if the group name ends in y or o, that terminal letter is changed to yl. The group name for a free radical always ends in yl. For example, $CH_3\cdot$ is methyl radical, and $CH_3{-}\ddot{\underset{..}{O}}\cdot$ is methoxyl radical. (*Chemical Abstracts* uses the group name only, without the separate word "radical." For complete clarity in general communication, inclusion of the class label is recommended.)

Alkyl radicals are classified as primary, secondary, and tertiary in the same way as alcohols are.

1. tert-Butyl radical is represented by the formula

____

and is classified as a __________ free radical.
(primary, etc.)

2. Benzyl radical is represented by the formula ____________________ and is classified as a __________ free radical.
(primary, etc.)

3. The free radical $CH_3{-}CH_2{-}\dot{C}H_2$ is named ____________________, and $CH_3{-}CH_2{-}\dot{C}H{-}CH_2{-}CH_3$ is named ____________________.

4. $(C_6H_5)_3C\cdot$ is named ____________________.

5. 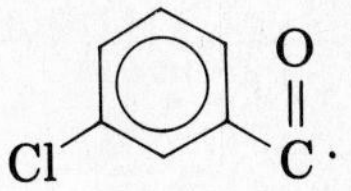 Cl O || C· is named ____________________, and

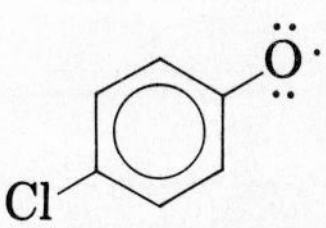

Cl Ö· is named ____________________.

6. The benzoyloxyl radical has the formula ____________________, and the dimethylaminyl radical has the formula ____________________.

7. The free radical

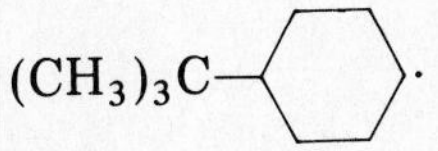

$(CH_3)_3C-$

is named ____________________ and classified as a __________ free radical.
(primary, etc.)

8. The radical

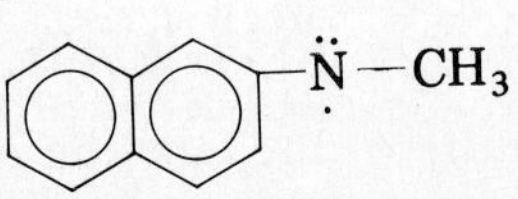

$\ddot{N}-CH_3$

is named ____________________.

Neutral particles that contain a divalent carbon (with six electrons) are called carbenes or methylenes. These structures may be named on the basis of the parent compound, $\ddot{C}H_2$, or by use of a systematic ending for a group with two free valences on the same carbon.

Carbene and methylene are synonymous names for the same parent compound, $\ddot{C}H_2$, and both are used. Preference for carbene as the parent compound name has been recently expressed by a Commission of IUPAC and earlier by a Nomenclature Committee of the American Chemical Society, and carbene connotes the reactive nature of the species in a way that methylene may not. *Chemical Abstracts*, however, uses methylene for $\ddot{C}H_2$ and does not use carbene. Substituted carbenes in which the electron-deficient carbon is not part of a ring are named in much the same way as is any other substituted parent compound.

Formulas for carbenes usually include two dots on the appropriate C to represent the unshared (nonbonding) electrons.

9. Dichlorocarbene has the formula ____________________, and phenylcarbene has the formula ____________________.

10. Hexyl radical contains __________ (number) carbons, and hexylcarbene contains __________ (number) carbons.

11. Hexyl radical is a two-word name, but hexylcarbene is a one-word name because ____.

12. The particle represented by the formula $CH_3{-}CH_2{-}CH_2{-}\ddot{C}H$ is named ______________________________, whereas the one represented by the formula $CH_3{-}CH_2{-}CH_2{-}\dot{C}H_2$ is named ______________________________.

13. Ethynylmethylene has the formula ______________________________, and dicyanomethylene has the formula ______________________________.

Carbenes in which the electron-deficient carbon is part of a ring may be named conveniently by use of the replacement name prefix, carbena (signifying replacement of carbon by carbene, in the same way that oxa signifies replacement of carbon by oxygen). Locant 1 is assigned to carbena, except in cyclic systems with fixed numbering. (This style has been recommended by the Committee on Nomenclature of the Division of Organic Chemistry of the American Chemical Society.) Alternatively, these (and acyclic) carbenes may be named by use of the systematic ending ylidene, for a group with two free valences on the same carbon. This ending replaces yl in the name of the corresponding free radical. (This style is used by *Chemical Abstracts*.) Recall that the bivalent group $-CH_2-CH_2-$ is called ethylene. Its isomer, $CH_3-CH<$, is called ethylidene. Ethylidene can be used as an acceptable name for methylcarbene and is used by *Chemical Abstracts*.

14. 4-tert-Butyl-1-carbenacyclohexane may be represented by the formula

____

and may also be named (*Chemical Abstracts* style) ____________________.

15. The particle

$$CH_2{=}CH-\triangleright:$$

may be named by a replacement name, ____________________,

or by a name with a systematic ending for a bivalent group,

____________________.

16. 2,4-Cyclopentadien-1-ylidene (*Chemical Abstracts* name) may be represented by the formula

____

and may also be named ____________________.
(replacement name)

Neutral particles that contain a monovalent nitrogen (with six electrons) are called nitrenes and are named in the same style as are carbenes. Nitrene is the name of the parent compound, $H\ddot{N}$:; therefore, the names of substituted nitrenes are one-word names.

17. Phenylnitrene is represented by the formula ____________________, and

cyanonitrene by the formula ____________________.

18. A substituent with a combination name is named toward its point of attachment. The substituent $CH_3O—CO—$ is named ______________, and the intermediate $CH_3—O—CO—\ddot{N}:$ is named ______________________________.

$$\begin{array}{l} \quad\quad\quad\quad\quad CH_3 \quad O \\ \quad\quad\quad\quad\quad\;\, | \quad\quad \| \\ CH_3—CH_2—CH——C—\ddot{N}: \end{array}$$

19. (structure above) is named ______________________________.

CATIONS AND ANIONS

Organic structures containing a carbon with a charge are called carbocations (+ charge) or carbanions (- charge). These ions can be formed (at least conceptually in all cases) from the corresponding free radicals by loss (cation) or gain (anion) of an electron. Although different styles of names for these ions have been accepted by IUPAC and used (sometimes abused) by chemists, the different styles are not equally consistent with other IUPAC rules of nomenclature or with the intended structures. The preferred approach to naming these ions is, as usual, the one that most clearly conveys the correct structural information and is, to our relief, the simplest one. For ions in which the charge is associated with a trivalent carbon, confusion about the intended structures is completely dispelled if the name of the corresponding free radical is modified merely by replacing radical by cation or anion, as appropriate. For example, $CH_3—CH_2—\dot{C}H_2$ is propyl radical, $CH_3—CH_2—\overset{+}{C}H_2$ is propyl cation, and $CH_3—CH_2—\overset{-}{C}H_2$ is propyl anion. Other ions, in which the charge is on a hetero atom (also with a valence one less than normal) are named from the corresponding free radical in the same way. This style is now recommended by a Commission of IUPAC. *Chemical Abstracts* does not use this style, however; it does not include the class labels in the names of the particles.

Other styles have led to faulty (or at least impaired) communication. IUPAC rules (revisions now under consideration) permit and *Chemical Abstracts* uses propylium for $CH_3—CH_2—\overset{+}{C}H_2$, but this style conflicts with the intended meaning of "plus a proton" for ium, may be taken (and used) to mean $CH_3—CH_2—\dot{C}H_2 + H^+$ or $CH_3—CH_2—\dot{C}H_3$, and should no longer be used. So much confusion has been associated with the use and misuse of "carbonium ion" in the names of specific carbocations that complete discontinuance of this term in specific names is strongly recommended.

Carbocations and carbanions are classified as primary, secondary, and tertiary on the same basis as are free radicals (and alcohols).

20. The carbocation $CH_3—CH_2—\overset{+}{C}H—CH_2—CH_3$ is named ______________________________ and is classified as a ______________________________ (primary, etc.) carbocation. The name 3-pentyl cation is incorrect because ______________________________.

21. The 3-nitrophenylmethyl anion is represented by the formula

____

and is classified as a ______________________________ carbanion.
(primary, etc.)

22. Allyl cation is represented by the formula ______________________________, and vinyl cation by the formula ______________________________.

23. The ion

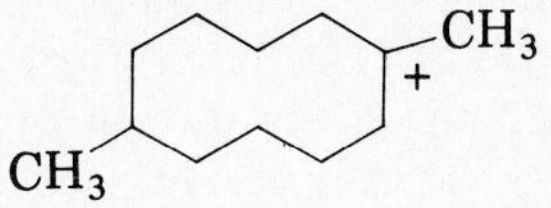

CH_3
+
CH_3

is named ______________________________

and is classified as a ______________________________ ion.
(primary, etc.)

24. 7-Methylbicyclo[2.2.1]hept-2-en-7-yl cation has the formula

______________________________.

25. The ion $(C_6H_5)_2C—CH_2—C_6H_5$ is named ______________________________

and is classified as a ______________________________ ion.
(primary, etc.)

26. The four species A, B, C, and D are named

A B C D

A ______________________________

B ______________________________

C ______________________________

D ______________________________

27. $Cl_3\bar{C}$ is named ____________________________________, and $CH_3-O-\overset{\overset{\displaystyle O}{\|}}{C}{}^{+}$ is named ____________________________________.

When carbocations and carbanions are to be identified as components in neutral, ionic compounds, the labels cation and anion are not included in the compound names. The suffix ide ("minus a proton") replaces the final e in the name of the appropriate neutral particle corresponding to the carbanion. For example, a $CH_3-\bar{C}H_2$ component of a neutral, ionic compound is named ethanide (ethane $-H^+$).

Appendix

SELECTED REFERENCES

Cahn, R. S., and O. C. Dermer, *An Introduction to Chemical Nomenclature*, 5th ed., Butterworths, London, Eng., 1979.

Chemical Abstracts, 9th Collective Index, 1972–76, Index Guide.

Fletcher, J. H., O. C. Dermer, and R. B. Fox, *Nomenclature of Organic Compounds* (*Advances in Chemistry Series* No. 126), American Chemical Society, Washington, D.C., 1974.

Hurd, C. D., "The General Philosophy of Organic Nomenclature," *Journal of Chemical Education*, **38**, 43 (1961).

IUPAC, *Nomenclature of Organic Chemistry*, *Sections A*, *B*, *C*, *D*, *E*, *F* and *H*, Pergamon Press, Oxford, Eng., 1979.

A BRIEF SUMMARY OF SOME KEY IUPAC RULES FOR SUBSTITUTIVE NAMES OF ORGANIC COMPOUNDS

1. The functional group to be used in the basis of the name must be identified. IUPAC rules include a full priority order of functional groups for names; a partial list is included in this Appendix. In general, a functional group with carbon in higher oxidation state (more bonds to hetero atoms) takes precedence over one with carbon in lower oxidation state.

2. The longest continuous chain of carbon atoms containing the functional group is the basis of the substitutive name. Atoms or groups other than hydrogen attached to the parent chain are called substitutents.

3. The parent compound is named by adding the appropriate systematic ending (suffix) to the name of the corresponding hydrocarbon; the final e in the hydrocarbon name is dropped for a suffix beginning with a vowel but is retained for a suffix beginning with a consonant.

4. Except for compounds containing multiple carbon-carbon bonds (which may be designated only by suffixes) in addition to other functional groups, no parent compound name may have more than one systematic ending.

5. The parent chain is numbered so that the functional group that is part of the parent compound is assigned the smaller possible number (locant). If numbering the parent chain in both directions gives the same locant to the functional group, the parent chain is numbered so that the smaller set of locants is used for the substituents. The smaller set of locants has the smaller locant at the first point of difference when alternative sets in sequence are compared term by term.

6. Locants for functional group(s) in the parent compound usually precede the hydrocarbon (stem) portion of the name. When the systematic ending clearly requires that the functional group include the terminal carbon of the parent chain (for example, -al for aldehyde), the locant 1 is omitted from the name.

7. A locant for each substituent must appear in the name even when the same locant must be used more than once for the same kind of substituent. The locant immediately precedes the substituent name to which it applies in the name of the compound.

8. Locants occurring together in the name are separated from each other by commas, and all locants are separated from the rest of the name by hyphens.

9. The carbon by which a hydrocarbon group substituent is attached to a parent chain is always designated position number 1 (of that group), but the locant 1 never appears in the name of the substituent group. Locants are used for substituents in the hydrocarbon group, even if the substituents are other hydrocarbon groups and are on position number 1. For example, sec-butyl, $CH_3—CH_2—\underset{|}{CH}—CH_3$, may also be named 1-methylpropyl (the systematic name and the only one used by *Chemical Abstracts*) but not 2-butyl.

SUBSTITUTIVE NAME PREFIXES AND SUFFIXES FOR SOME IMPORTANT FUNCTIONAL GROUPS

(Arranged in Descending Order of Preference for Citation as Suffixes)

Class	Formula of Group[a]	Prefix[b]	Suffix[c]
Carboxylic acids	—COOH	carboxy	-carboxylic acid
	—(C)OOH	—	-oic acid

SUBSTITUTIVE NAME PREFIXES AND SUFFIXES FOR SOME IMPORTANT FUNCTIONAL GROUPS *(Continued)*

Class	Formula of Group[a]	Prefix[b]	Suffix[c]
Sulfonic acids	$—SO_3H$	sulfo	-sulfonic acid
Esters	—COOR	R[d]-oxycarbonyl	R[d] . . .-carboxylate
	—(C)OOR	—	R[d] . . .-oate
Acid halides	—CO—Hal	haloformyl or halocarbonyl	-carbonyl halide
	—(C)O—Hal	—	-oyl (or -yl) halide
Amides	$—CO—NH_2$	carbamoyl	-carboxamide
	$—(C)O—NH_2$	—	-amide
Nitriles	—C≡N	cyano	-carbonitrile
	—(C)≡N	—	-nitrile
Aldehydes	—CHO	formyl	-carbaldehyde[e]
	—(C)HO	oxo	-al
Ketones	>(C)=O	oxo	-one
Alcohols	—OH	hydroxy	-ol
Phenols	—OH	hydroxy	-ol
Thiols	—SH	mercapto	-thiol
Amines	$—NH_2$	amino	-amine
Ethers	—OR	R[d]-oxy	—
Sulfides	—SR	R[d]-thio	—

[a] C in parentheses is included in the stem of the parent chain and not in the prefix or suffix.

[b] Functional group is treated as a substituent.

[c] Functional group is part of parent compound; suffix is added to name of corresponding hydrocarbon.

[d] R is alkyl, aryl, etc.; when R is part of a prefix, the name of the R group is written as part of the prefix name without a hyphen.

[e] *Chemical Abstracts* uses -carboxaldehyde in indexes.

NAMES OF SOME IMPORTANT PARENT COMPOUNDS NOT SPECIFICALLY INCLUDED IN THIS PROGRAM

The rules of nomenclature included in this program are readily applied to substituted compounds based on the following parent compounds. Fixed numbering in these parent compounds is indicated by the locants included with the formulas.

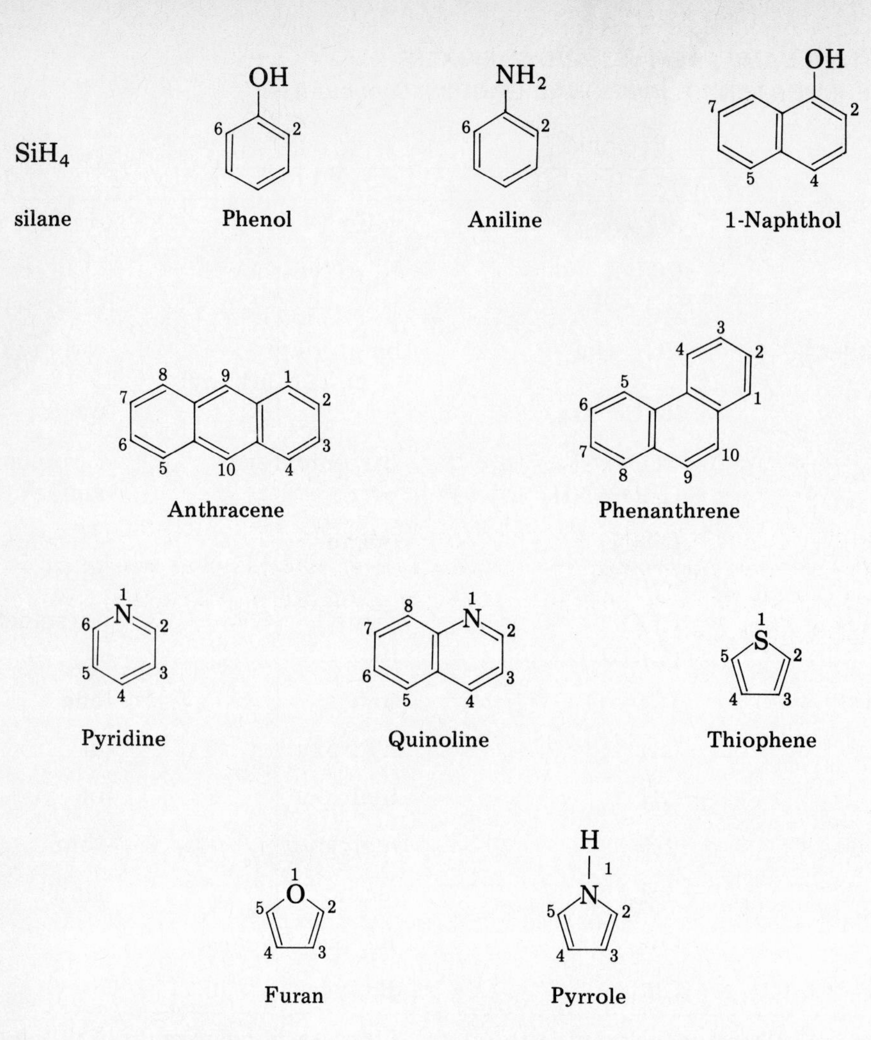

REPLACEMENT NAME PREFIXES FOR HETERO ATOMS[a]

(Listed in Descending Order of Precedence)

Element	Prefix
oxygen	oxa
sulfur	thia
nitrogen	aza
phosphorus	phospha
silicon	sila
boron	bora

[a] For a more complete list, see either the IUPAC or Fletcher entry in the Selected References.

Answer Sheets

IMPORTANT NOTE

Remove the answer sheet for the chapter being studied by tearing it from the book along the perforated line. Cover the answers with an index card or a sheet of paper, and after you have written your answer in the appropriate blank expose the answers one at a time to check your response.

CHAPTER 1 ALKANES

1. 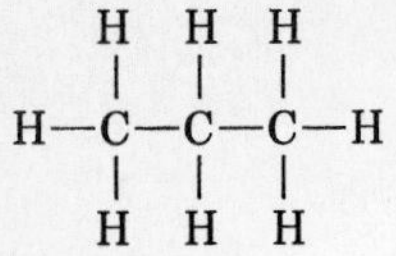 $$\begin{array}{ccccccc} & H & & H & & H & \\ & | & & | & & | & \\ H- & C & - & C & - & C & -H \\ & | & & | & & | & \\ & H & & H & & H & \end{array}$$

2. four
   one

3. $CH_3-CH_2-CH_3$

4. four
   four

5. meth
   ane

6. ethane
   propane

7. $CH_3-CH_2-CH_2-CH_3$

   $$\left(\text{or } \begin{array}{ccccc} & & CH_3 & & \\ & & | & & \\ CH_3 & - & CH & - & CH_3 \end{array}\right)$$

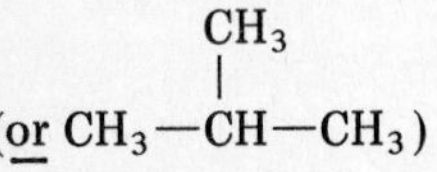

8. alkane

9. seven

10. 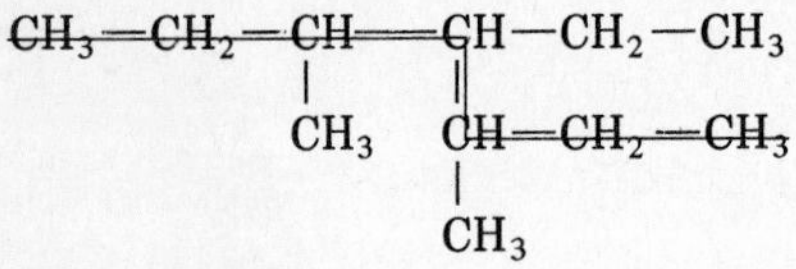 $$\begin{array}{ccccccccccc} CH_3 & - & CH_2 & - & CH & - & CH & - & CH_2 & - & CH_3 \\ & & & & | & & | & & & & \\ & & & & CH_3 & & CH & - & CH_2 & - & CH_3 \\ & & & & & & | & & & & \\ & & & & & & CH_3 & & & & \end{array}$$

11. hept
    heptane

12. three

13. meth
    yl
    two
    ethyl
    alkyl

14. methyl
    methyl
    ethyl

15. dimethyl

16. ethyldimethylheptane

17. eleven
    seven
    two, one, one
    (or one, one, two)
    eleven

18. 3, 4, 5

19. 4-ethyl-3,5-dimethylheptane

20. the longest continuous chain of carbon atoms, or the parent chain, or the chain of carbon atoms serving as a basis of the name
    substituents
    three
    positions of substituents along the parent (longest continuous) chain
    commas
    hyphens

21. longest continuous
    seven
    hept
    heptane
    three
    methyl, ethyl, ethyl

22. 2, 4, 5
    3, 4, 6
    2, 4, 5
    2; 4 and 5
    commas; hyphens

23. 4,5-diethyl-2-methylheptane

24. five
    pentane
    2,2,4-trimethylpentane

25. 5,5-diethyldecane

26. 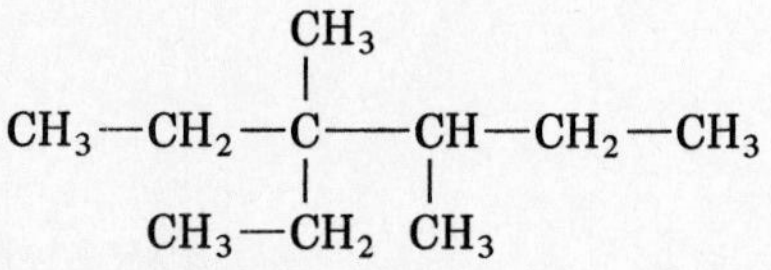 $$\begin{array}{ccccccccc} & & & & CH_3 & & & & \\ & & & & | & & & & \\ CH_3 & - & CH_2 & - & C & - & CH & - & CH_2-CH_3 \\ & & & & | & & | & & \\ & & & & CH_3-CH_2 & & CH_3 & & \end{array}$$

27. 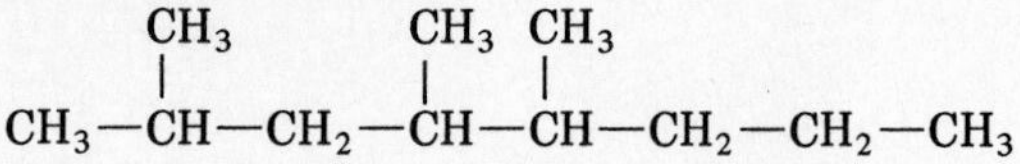 $$\begin{array}{ccccccccccccccc} & & CH_3 & & & & CH_3 & & CH_3 & & & & & & \\ & & | & & & & | & & | & & & & & & \\ CH_3 & - & CH & - & CH_2 & - & CH & - & CH & - & CH_2 & - & CH_2 & - & CH_3 \end{array}$$

28. C_6H_{14}
    C_6H_{14}
    isomers

29. $CH_3-CH_2-CH_2-CH_2-CH_3$

$$\begin{array}{l}CH_3\\ |\\ CH_3-CH-CH_2-CH_3\end{array}$$

$$\begin{array}{l}CH_3\\ |\\ CH_3-C-CH_3\\ |\\ CH_3\end{array}$$

30. four
one

31. pentane
2-methylbutane
2,2-dimethylpropane (locants unnecessary in the last two names because substituent methyls cannot be on any other carbon, but *Chemical Abstracts* uses them)

32. $CH_3-CH_2-CH_2-CH_3$

$$\begin{array}{l}CH_3\\ |\\ CH_3-CH-CH_3\end{array}$$

2-methylpropane (locant unnecessary, but used by *Chemical Abstracts*; see last response in item 31)

33. pent
5
iso
isopentane

34. hexane
isohexane
2-methylpentane

35. 3-methylpentane

36. $CH_3-CH_2-CH_2-CH_2-CH_3$

$$\begin{array}{l}CH_3\\ |\\ CH_3-CH_2-CH-CH_3\end{array}$$

37.
$$\begin{array}{l}CH_3\\ |\\ CH_3-C-CH_3\\ |\\ CH_3\end{array}$$

38. neopentane

39. 2,2-dimethylbutane

40. 3,3-dimethylhexane

41. C_4H_8

$$\begin{array}{l}CH_2-CH_2\\ ||\\ CH_2-CH_2\end{array}$$

cyclobutane

42. cyclohexane

43. ethylcyclopentane

44. ethyl
1
3
1-ethyl-3-methylcyclohexane

45. 1,3-dimethylcyclobutane

46. ethyl
1,2,4,7
1,2,3,5
2-ethyl-1,3,5-trimethylcycloheptane

1. methyl
   ethyl

2. 5
   3
   1,1,4
   1,1-diethyl-4-methylpentyl

3. 1-ethylbutyl
   1-ethylbutylcycloheptane

4. 1-methylbutyl
   2-methylbutyl
   2,2-dimethylpropyl

5. 1,1,3-trimethylpentyl
   1-methylpropyl
   1-methylpropyl
   1
   4
   1-(1-methylpropyl)-4-(1,1,3-trimethylpentyl)cyclooctane

6. but
   butyl
   primary

7. secondary

8. primary
   tertiary

9. primary

10. $CH_3-CH_2-\underset{|}{C}H-CH_3$

11. $CH_3-\overset{\overset{\displaystyle CH_3}{|}}{\underset{\underset{\displaystyle CH_3}{|}}{C}}-$
    tert-butyl

12. *tert*-pentyl
    1,1-dimethylpropyl

13. $CH_3-\overset{\overset{\displaystyle CH_3}{|}}{C}H-CH_3$
    $CH_3-\overset{\overset{\displaystyle CH_3}{|}}{C}H-CH_2-$
    primary

14. $CH_3-\overset{\overset{\displaystyle CH_3}{|}}{C}H-CH_2-CH_3$
    $CH_3-\overset{\overset{\displaystyle CH_3}{|}}{C}H-CH_2-CH_2-$
    primary

15. isohexyl

16. $CH_3-\overset{\overset{\displaystyle CH_3}{|}}{\underset{\underset{\displaystyle CH_3}{|}}{C}}-CH_3$
    $CH_3-\overset{\overset{\displaystyle CH_3}{|}}{\underset{\underset{\displaystyle CH_3}{|}}{C}}-CH_2-$

17. branched

18. $CH_3-CH_2-\overset{|}{C}H-CH_3$

19. $CH_3-\underset{\underset{\displaystyle CH_3}{|}}{C}H-CH_2-$

20. primary
    secondary

21. 1-methylpropyl
    2-methylpropyl
    1,1-dimethylethyl

22. secondary
    isopropyl

23. isopropyl
    isopropylcyclobutane

24. 10
    methyl and *sec*-butyl
    2 and 5
    sec-butyl
    5-*sec*-butyl-2-methyldecane

25. 10
    isobutyl
    5
    decane
    5-isobutyldecane

26. isopropyl
isopropyl chloride
2-chloropropane

27. $CH_3{-}CH_2{-}CH_2{-}CH_2{-}Cl$

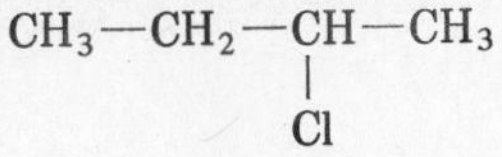

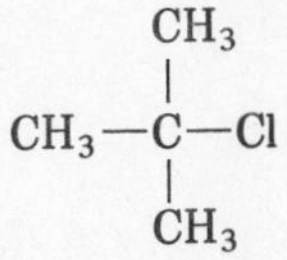

```
CH3—CH2—CH—CH3
         |
         Cl
```

```
     CH3
      |
CH3—C—Cl
      |
     CH3
```

28. 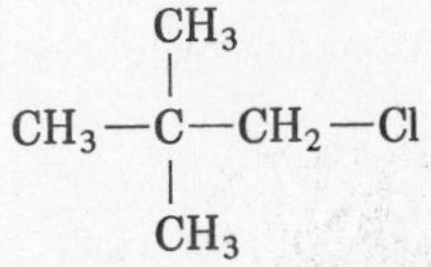
```
     CH3
      |
CH3—C—CH2—Cl
      |
     CH3
```

1-chloro-2,2-dimethylpropane
primary

29. 1-chloro-3-methylbutane
isopentyl chloride
primary

30. pentane
5
C—C—C—C—C
3
2, 2, and 4

```
1  2  3  4  5
C—C—C—C—C
```

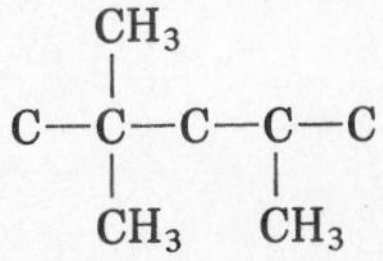

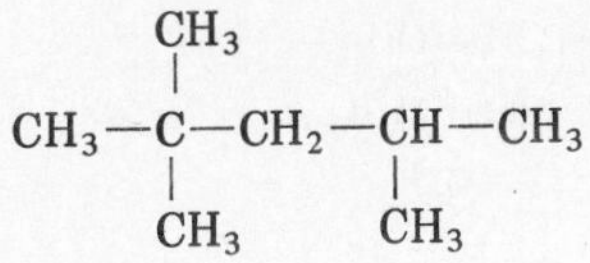

```
   CH3
    |
C—C—C—C—C
    |     |
   CH3   CH3
```

```
     CH3
      |
CH3—C—CH2—CH—CH3
      |         |
     CH3       CH3
```

31. 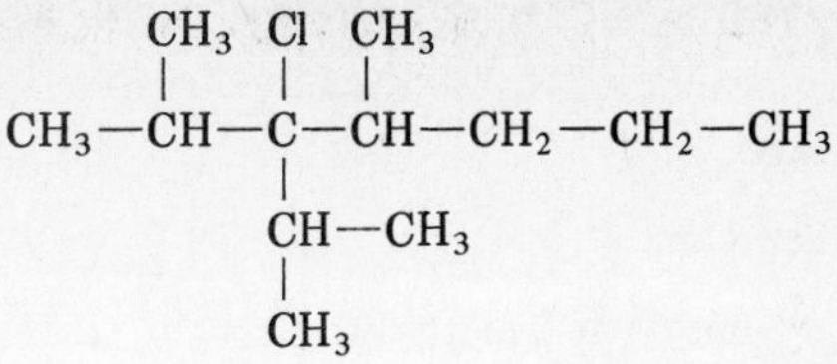
```
CH3  Cl  CH3
 |    |   |
CH3—CH—C—CH—CH2—CH2—CH3
          |
         CH—CH3
          |
         CH3
```

32. 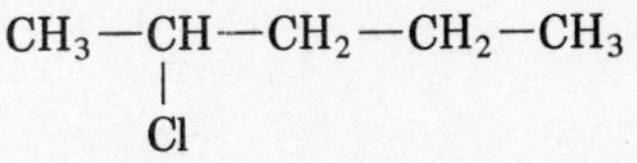
```
CH3—CH—CH2—CH2—CH3
     |
     Cl
```

```
CH3—CH2—CH—CH2—CH3
         |
         Cl
```

```
CH3—CH—CH—CH3
     |   |
     Cl  CH3
```

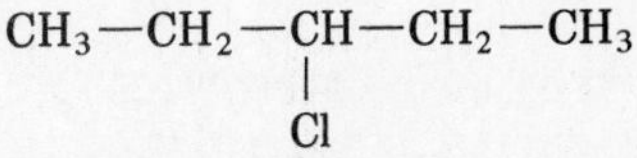

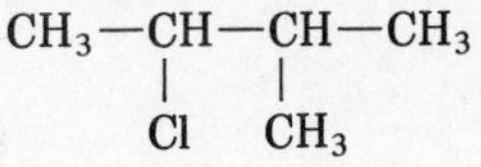

33. 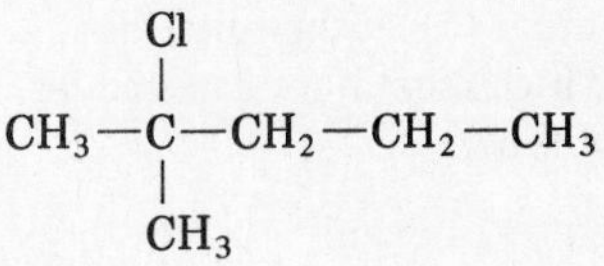
```
     Cl
      |
CH3—C—CH2—CH2—CH3
      |
     CH3
```

```
          Cl
           |
CH3—CH2—C—CH2—CH3
           |
          CH3
```

```
     Cl  CH3
      |   |
CH3—C—CH—CH3
      |
     CH3
```

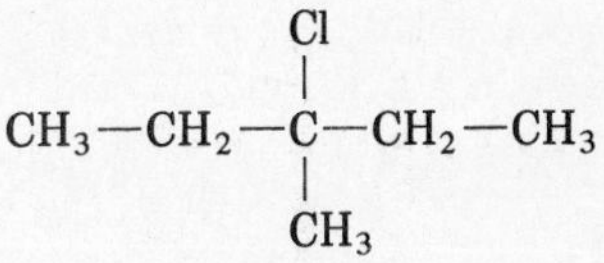

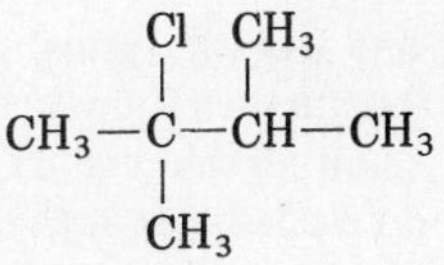

34. 2-chloro-3-methylbutane

35. 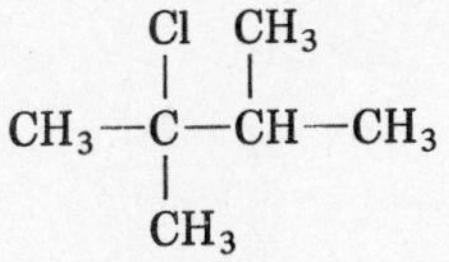
```
     Cl  CH3
      |   |
CH3—C—CH—CH3
      |
     CH3
```

CHAPTER 3 NOMENCLATURE OF ALKENES

1. seven

2. heptene

3. sec-butyl (or 1-methylpropyl)

4. 2
   3

5. 2-heptene

6. 3-sec-butyl-2-heptene [or 3-(1-methylpropyl)-2-heptene]

7. hexene

8. 1
   3, 5, 5

9. that numbering would incorrectly assign a higher number (5) to the functional group (C=C), which must have the smaller possible number

10. 3,5,5-trimethyl-1-hexene

11. $CH_3-CH_2-CH=CH_2$, $CH_3-CH=CH-CH_3$, and $CH_3-C(CH_3)=CH_2$

12. 1-butene, 2-butene, and 2-methylpropene (In the last name the number is unnecessary because the substituent methyl cannot be on any other carbon if the basis of the name is to be propene).

13. cyclooctene

14.

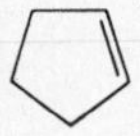

15. 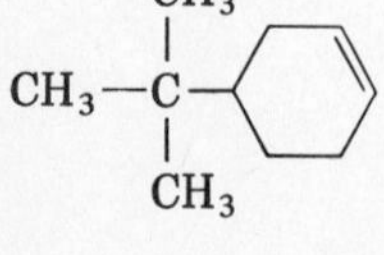 $CH_3-C(CH_3)_2-$

16. 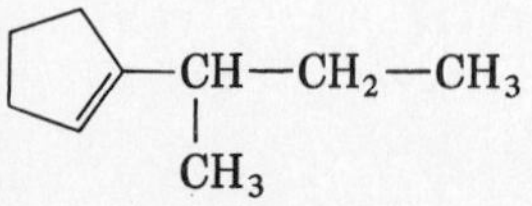 $-CH(CH_3)-CH_2-CH_3$

17. 3
    3-chloropropene (or 3-chloro-1-propene)

18. 1 and 2
    1-chloro-2-isobutylcyclobutene

19. 1 and 5
    2 and 3
    1 and 5
    1-chloro-5-methylcyclopentene

20. 3,4,4,5,6 and 3,4,5,5,6
    3,4,4,5,6
    3,5-dichloro-4,4,6-trimethylcyclohexene

21. 1-propenyl

22. $CH_2=CH-$

23. $CH_2=CH-Cl$

24. $CH_2=CH-CH_2-Cl$

25. primary

26. vinyl
    4
    4-vinylcyclohexene

27. allyl bromide

28. 2-methyl-2-propenyl

29. 1-methyl-1-propenyl

30. 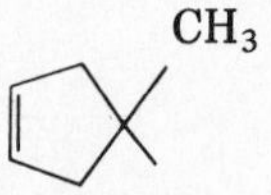 CH_3

    tertiary

31. $CH_3-CH_2-CH=CH_2$, $CH_3-CH=CH-CH_3$, and $CH_3-C(CH_3)=CH_2$

32. one
    2-butene

33. 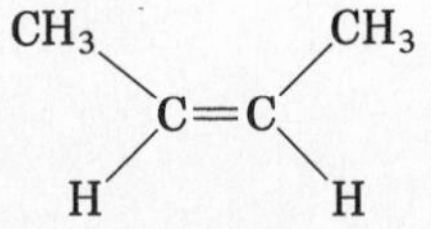 $(CH_3)HC=CH(CH_3)$ (both CH_3 on the same side) and $(CH_3)HC=CH(CH_3)$ (the two CH_3 on opposite sides)

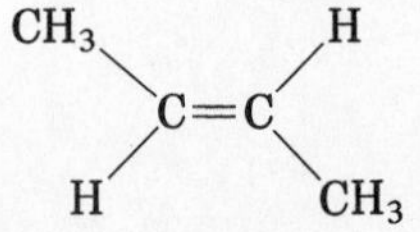

34. 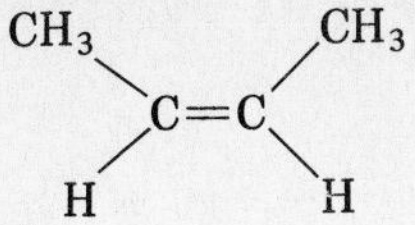 CH_3 CH_3
$C=C$
H H

35. 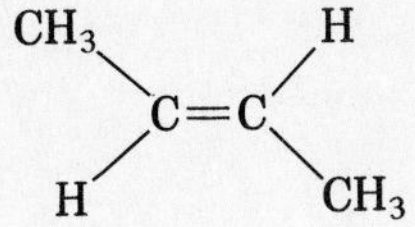 CH_3 H
$C=C$
H CH_3

36. 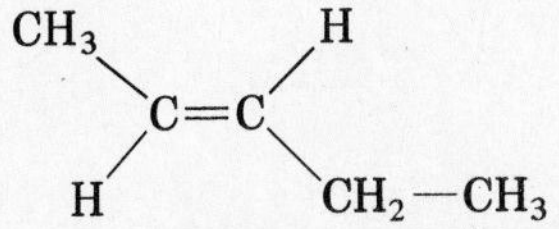 CH_3 H
$C=C$
H CH_2-CH_3

37. the parent chain extends from opposite sides of the alkene linkage

38. 5-methyl-cis-2-hexene

43. C, H, H,
C, C, H
H, H, H
lower
higher
E
(E)-4-ethyl-3,5-dimethyl-3-heptene
trans

44. E
cis

45. Cl
$C=C$
CH_3-CH_2 H

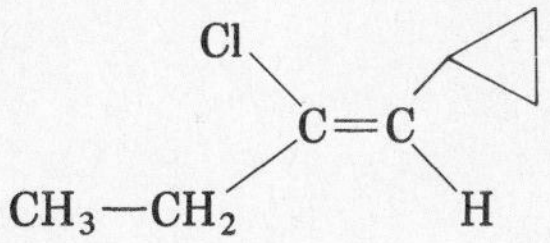

46. E

---

39. 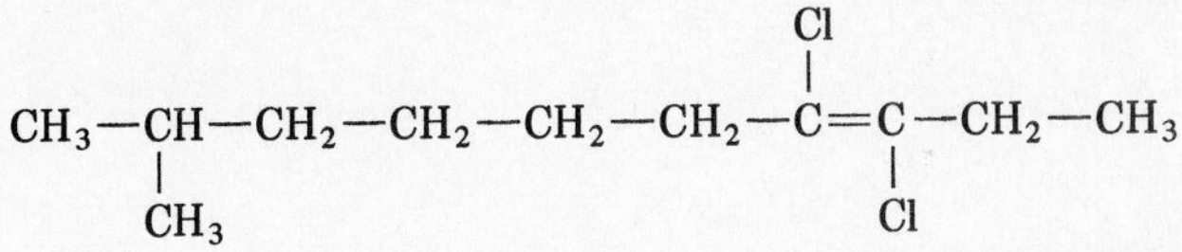 $CH_3-CH(CH_3)-CH_2-CH_2-CH_2-CH_2-C(Cl)=C(Cl)-CH_2-CH_3$

40.

or

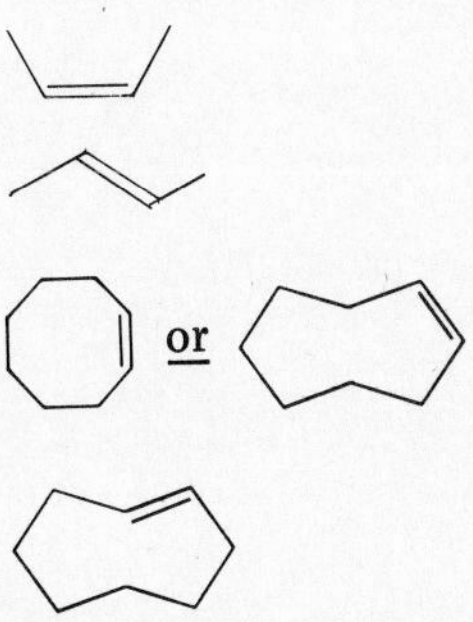

41. trans
10
trans-cyclodecene

42. Cl, H
higher atomic number gives higher priority (rule 1)
higher atomic numbers of second atoms out give higher priority (rule 3)
CH_3-CH_2-, CH_3-
opposite
E
1-butene
(E)-1-chloro-2-methyl-1-butene

47. $CH_2=CH-CH_2-CH=CH_2$
$CH_2=CH-CH=CH-CH_3$

48. 1,3-butadiene

49. 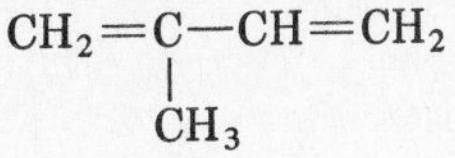 $CH_2=C(CH_3)-CH=CH_2$

50. $CH_2=C=CH_2$

51. 1,2-butadiene

52.

53. conjugated

54. isolated

55. $CH_2=C=CH_2$
cumulated

56. 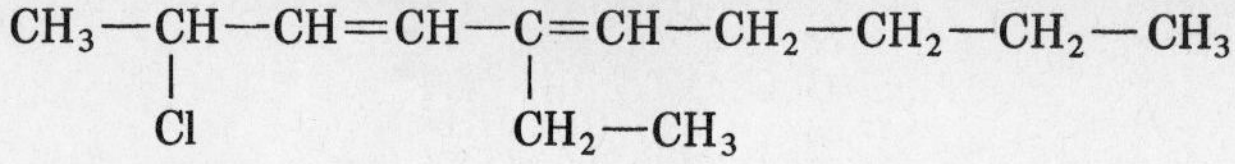 $CH_3-\underset{\displaystyle Cl}{\underset{|}{C}H}-CH=CH-\underset{\displaystyle CH_2-CH_3}{\underset{|}{C}}=CH-CH_2-CH_2-CH_2-CH_3$

conjugated

CH_3-CH_2 H

$CH_3-CH-Cl$ C=C

C=C $CH_2-CH_2-CH_2-CH_3$

H H

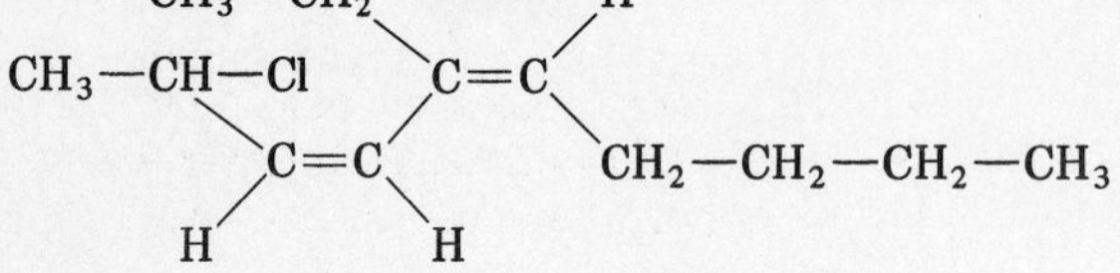

57.

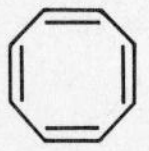

58. 1,2-dibromoethane

CHAPTER 4 ALKYNES

1. propyne
2. 7
   hept
   yne
   heptyne
   3
   3-heptyne
   methyl, ethyl
   2, 5
   5-ethyl-2-methyl-3-heptyne
3. 2-butyne
4. $CH_3{-}CH_2{-}C{\equiv}C{-}CH_2{-}CH_3$
5. 8
   oct
   $C{\equiv}C$
   $C{-}C{\equiv}C{-}C{-}C{-}C{-}C{-}C$
   Cl

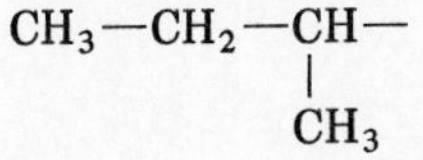

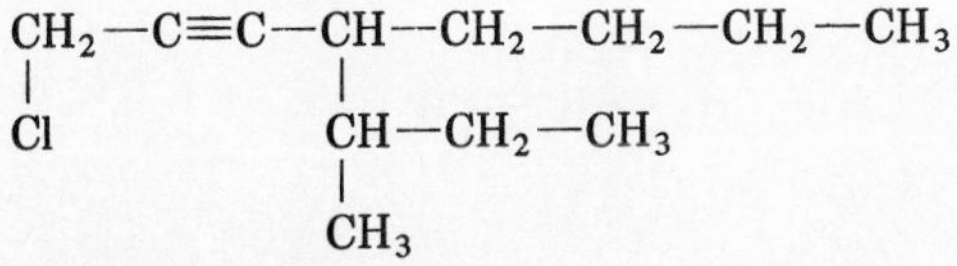

$$\begin{array}{l} CH_3{-}CH_2{-}CH{-} \\ \qquad\qquad\quad | \\ \qquad\qquad\; CH_3 \end{array}$$

$$\begin{array}{lllllll} CH_2 & {-}C{\equiv}C{-} & CH & {-}CH_2 & {-}CH_2 & {-}CH_2 & {-}CH_3 \\ | & & | & & & & \\ Cl & & CH{-}CH_2{-}CH_3 & & & & \\ & & | & & & & \\ & & CH_3 & & & & \end{array}$$

   (1-methylpropyl)
   1-chloro-4-(1-methylpropyl)-2-octyne

6. 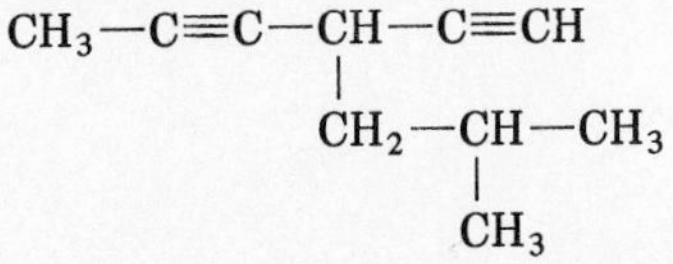
$$\begin{array}{l} CH_3{-}C{\equiv}C{-}CH{-}C{\equiv}CH \\ \qquad\qquad\quad\; | \\ \qquad\qquad\; CH_2{-}CH{-}CH_3 \\ \qquad\qquad\qquad\qquad | \\ \qquad\qquad\qquad\; CH_3 \end{array}$$

   3-(2-methylpropyl)-1,4-hexadiyne

7. 1,6-cyclodecadiyne
   3,4,9
   3,4,9-trimethyl-1-6-cyclodecadiyne
8. butenyne (locants are unnecessary, because only one butenyne is possible, but *Chemical Abstracts* uses them: 1-buten-3-yne)
9. 4-chloro-7-isopropyl-6-nonen-2-yne
   6-chloro-3-isopropyl-2-nonen-7-yne
10. 4-ethyl-4-methyl-1-hexen-5-yne
11. ethynyl
12. 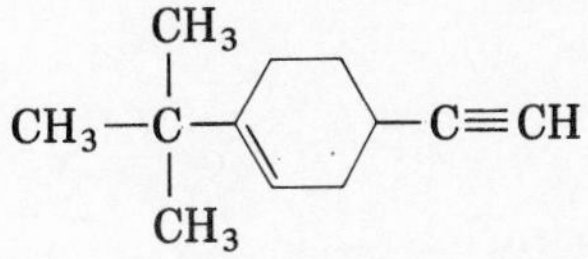
$$\begin{array}{l} \qquad\;\; CH_3 \\ \qquad\quad\;\; | \\ CH_3{-}C{-}\langle\text{cyclohexene ring}\rangle{-}C{\equiv}CH \\ \qquad\quad\;\; | \\ \qquad\;\; CH_3 \end{array}$$

13. $HC{\equiv}C{-}CH_2{-}$
    $CH_3{-}C{\equiv}C{-}$
14. $-CH_2{-}C{\equiv}C{-}CH{=}CH_2$,

$$\begin{array}{l} \qquad\qquad\qquad\;\; | \\ CH_3{-}C{\equiv}C{-}C{=}CH_2, \text{ and} \end{array}$$

   $CH_3{-}C{\equiv}C{-}CH{=}CH{-}$
   4-penten-2-ynyl, (1-propynyl)ethenyl, and 1-penten-3-ynyl

1. ethylbenzene
2. H—C≡C—

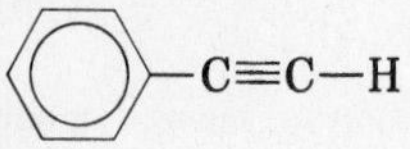

   —C≡C—H

3. 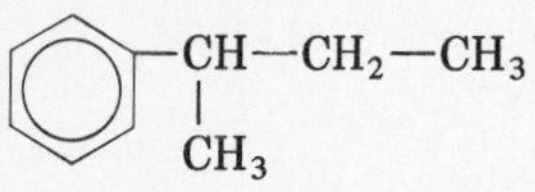 $-CH-CH_2-CH_3$ (with CH_3 on the CH)

4. neopentyl
   neopentylbenzene
5. triphenylmethane
6. 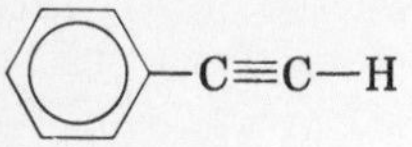 —C≡C—H

   phenylethyne
7. 2-nonene
   4
   6
   4-methyl-6-phenyl-2-nonene

12. 1,2,4
    2-isopropyl-1,4-dimethylbenzene
13. CH_3, CH_3, CH_3
14. $-CH_3$, $-CH_3$

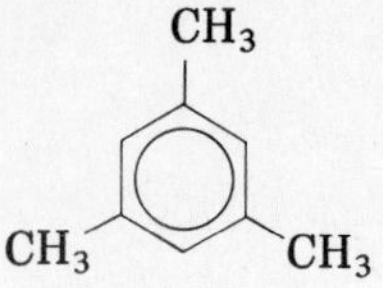

    CH_3, CH_3

    CH_3- $-CH_3$
15. p-diisopropylbenzene

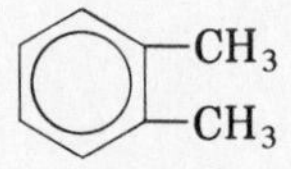

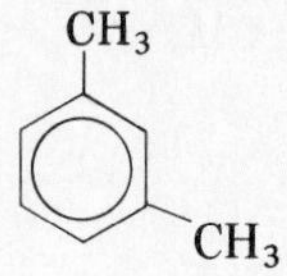

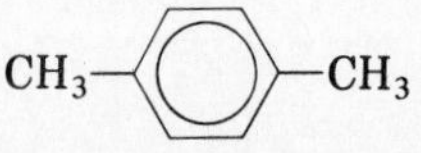

8. 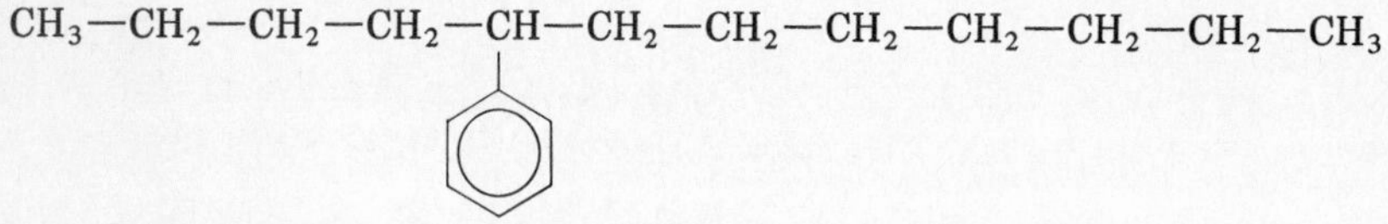 $CH_3-CH_2-CH_2-CH_2-CH-CH_2-CH_2-CH_2-CH_2-CH_2-CH_2-CH_3$

   or $CH_3-(CH_2)_3-CH-(CH_2)_6-CH_3$ (with ϕ on the CH)

9. 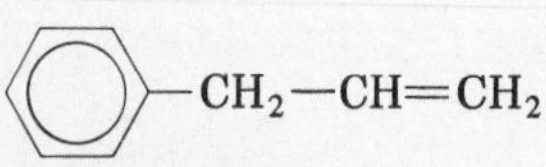 $-CH_2-CH=CH_2$

   or $C_6H_5-CH_2-CH=CH_2$
   or $\phi-CH_2-CH=CH_2$
   3-phenylpropene
   $C_6H_5-CH=CH-CH_3$
   1-phenylpropene
10. 1, 3
11. 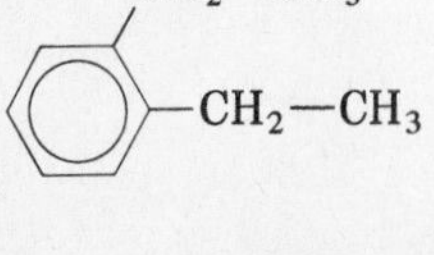 CH_2-CH_3

    $-CH_2-CH_3$

16. o (ortho)
17. 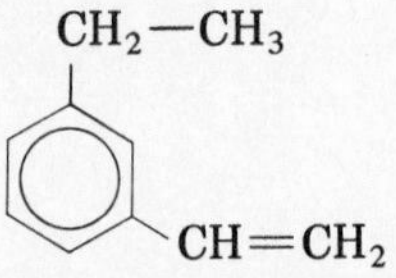 CH_2-CH_3

    $CH=CH_2$
18. o-hexylisobutylbenzene

    CH_2-CH_3

19. 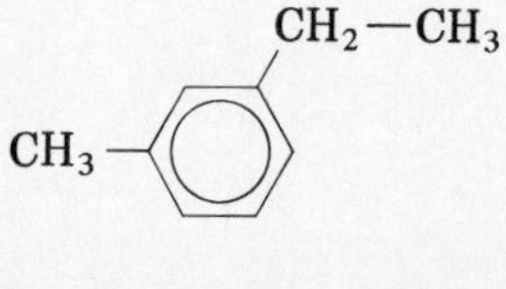 CH_3-

20. 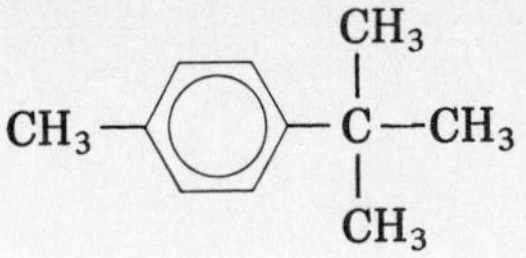 CH_3—(benzene ring)—$C(CH_3)_3$

CH_3

CH_3—⟨ring⟩—C—CH_3

CH_3

21. 2-allyltoluene *or* *o*-allyltoluene
or 2-(2-propenyl)toluene
or *o*-(2-propenyl)toluene

22. 1-*sec*-butyl-4-methylbenzene *or*
1-methyl-4-(1-methylpropyl)benzene
4-*sec*-butyltoluene *or* *p*-*sec*-butyltoluene

23. 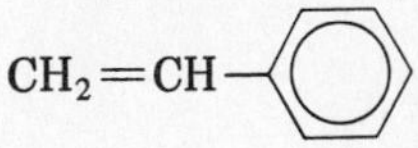 $CH_2{=}CH$—⟨ring⟩ or $CH_2{=}CH{-}\phi$

24. 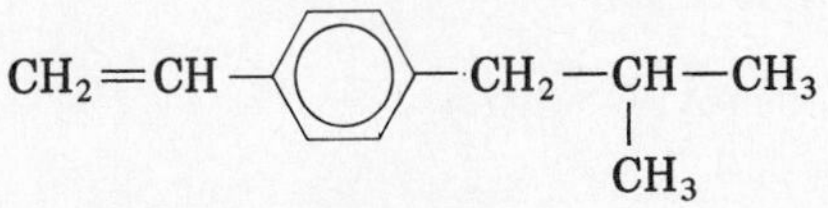 $CH_2{=}CH$—⟨ring⟩—$CH_2{-}CH(CH_3){-}CH_3$

25. (1-methylethenyl)benzene

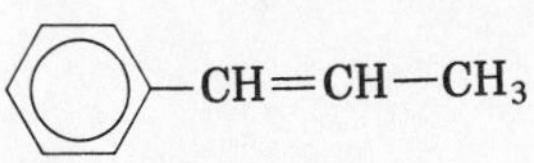

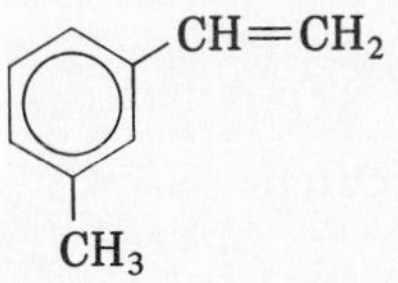

⟨ring⟩—$CH{=}CH{-}CH_3$

$CH{=}CH_2$

CH_3

26. 2-phenyl-1-pentene

27.

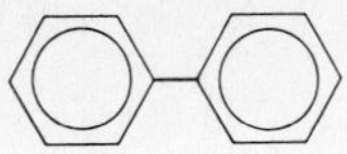

28. 2-*tert*-butyl-2′,6-dimethylbiphenyl

29. 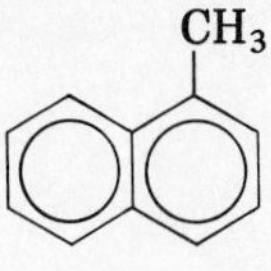 CH_3

CH_3 CH_3

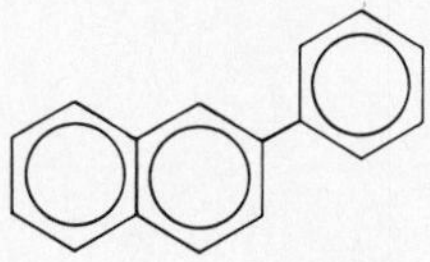

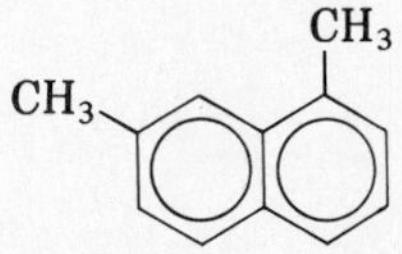

30. 2-methyl-4-(2-naphthyl)-1-butene
2-(3-methyl-3-butenyl)naphthalene

31. 1,4
1,4-dihydronaphthalene
1,4,5,8
1,4,5,8-tetrahydronaphthalene
1,2,3,4-tetrahydronaphthalene

CHAPTER 6 ALCOHOLS

1. isopropyl alcohol
2. $CH_3-\underset{OH}{\underset{|}{CH}}-CH_2-CH_3$
3. alcohol is a class name rather than the name of a specific compound
4. isopentyl
   isopentyl alcohol
5. allyl
   allyl alcohol
6. $CH_3-\overset{CH_3}{\overset{|}{\underset{CH_3}{\underset{|}{C}}}}-OH$
7. CH_3-CH_2-OH
   primary
8. $CH_3-\underset{CH_3}{\underset{|}{CH}}-CH_2-OH$
   primary
9. primary
10. the OH group is attached to a primary carbon, that is, to a carbon that is attached to only one other carbon
11. secondary
12. $CH_3-\underset{OH}{\underset{|}{CH}}-CH_2-CH_3$
    sec-butyl alcohol
13. $CH_3-\overset{OH}{\overset{|}{CH}}-CH_2-CH_2-CH_3,$
    $CH_3-CH_2-\overset{OH}{\overset{|}{CH}}-CH_2-CH_3,$
    $CH_3-\underset{CH_3}{\underset{|}{CH}}-\overset{OH}{\overset{|}{CH}}-CH_3$
14. $CH_3-\overset{CH_3}{\overset{|}{\underset{CH_3}{\underset{|}{C}}}}-OH$
    tert-butyl alcohol
15. primary
16. 3-pentanol
17. $CH_3-\overset{OH}{\overset{|}{CH}}-CH_2-CH_3$
18. cyclohexanol
19. 3-methyl-2-butanol
20. 9
    nonane
21. nonanol
22. 4
    4-nonanol
23. 3
    chloro, phenyl, and isopropyl
    3, 4, and 6
24. hyphens
    commas
25. 3-chloro-6-isopropyl-4-phenyl-4-nonanol
26. tertiary
27. $C_6H_5-CH_2-OH$
    phenylmethanol
28. $CH_3-\underset{OH}{\underset{|}{CH}}-CH_2-OH$
29. 1,3-propanediol
30. 1,2,3-propanetriol
31. 2 and 4
    lower possible
    5,6-dimethyl-6-phenyl-2,4-heptanediol

32. 1,2,4-cyclohexanetriol

33. $CH_2{=}CH-CH_2-OH$
2-propen-1-ol

34. 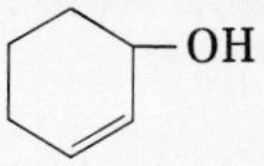 —OH

35. 4-cycloocten-1-ol

36. $CH_3-CH_2-\underset{\displaystyle OH}{\overset{\displaystyle CH_3}{\overset{|}{\underset{|}{C}}}}-C{\equiv}CH$

37. 7
heptenol
3, 5
5-hepten-3-ol
2
methyl, ethyl
6, 4
4-ethyl-6-methyl-5-hepten-3-ol

38. $CH_3-\underset{\displaystyle OH}{\underset{|}{CH}}-\underset{\displaystyle C_6H_5}{\underset{|}{C}}{=}CH-\underset{\displaystyle Cl}{\underset{|}{CH}}-CH_3$

39. 6-methyl-3-heptene-2,5-diol

40. 4-sec-butyl-4-cyclohexene-1,2-diol

41. 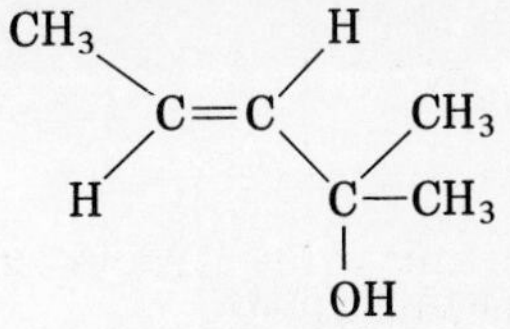 CH_3, H, C=C, H, CH_3, C—CH_3, OH

42. 2-methyl-6-phenyl-cis-3-decen-1-ol

43. 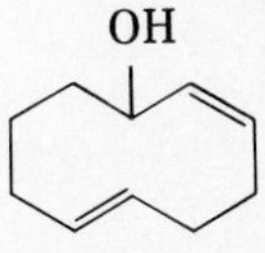 OH

44. 9
2,6,6-trimethyl-1-cyclohexenyl trans
3,7-dimethyl-9-(2,6,6-trimethyl-1-cyclohexenyl)-trans-2-trans-4-trans-6-trans-8-nonatetraen-1-ol

45. $HO-CH_2-\underset{\displaystyle CH_2-OH}{\underset{|}{CH}}-CH_2-OH$

46. 4-hydroxyphenyl
5-(4-hydroxyphenyl)-2-heptanol

CHAPTER 7 ETHERS

1. ethoxyethane

2. 7
   heptane
   2, 3, 5
   ethoxy
   5-ethoxy-2,3-dimethylheptane

3. CH_3-O- ⟨benzene ring⟩ $-O-CH_3$

   $CH_3-CH_2-CH_2-CH(O-CH_2-CH_3)_2$

4. phenoxybenzene

5. heptane
   4
   2, 4, 6
   4-chloro-6-cyclohexyloxy-2,2-dimethylheptane

6. 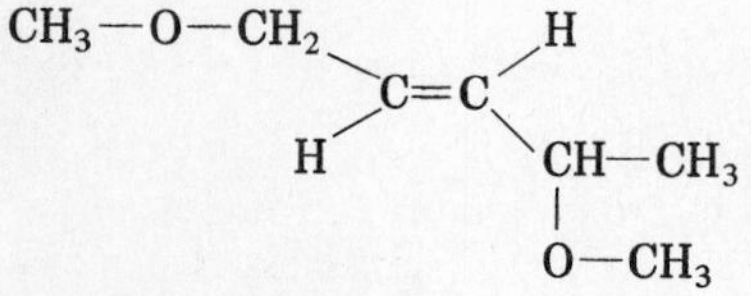 CH_3-O-CH_2 and H on one carbon, H and $CH-CH_3$ on the other of $C=C$; the CH bears $O-CH_3$

   $$\begin{array}{ccc} CH_3-O-CH_2 & & H \\ & C=C & \\ H & & CH-CH_3 \\ & & | \\ & & O-CH_3 \end{array}$$

7. cyclohexanol
   ethoxy, phenyl
   1, 5
   5-ethoxy-2-phenylcyclohexanol

8. 10
   decyne
   2
   isopropoxy
   5
   5-isopropoxy-2-decyne

9. $C_6H_5-O-C_6H_5$ or ⟨benzene ring⟩$-O-$⟨benzene ring⟩

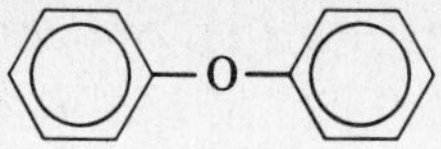

   $$\begin{array}{l} CH_3-CH-O-C_6H_5 \\ \quad\;\; | \\ \quad\; CH_3 \end{array}$$

10. allyl
    diallyl ether

11. 2-sec-butoxybutane

12. tert-pentyl
    cyclopropyl
    cyclopropyl tert-pentyl ether
    2-cyclopropoxy-2-methylbutane

13. 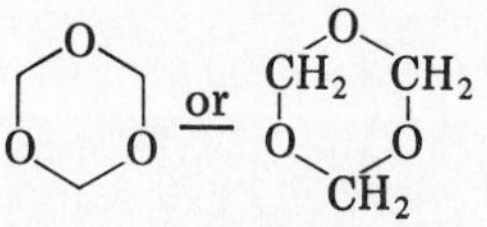 (six-membered ring with O at positions 1, 3, 5) or

    $$\begin{array}{ccc} & O & \\ CH_2 & & CH_2 \\ | & & | \\ O & & O \\ & CH_2 & \end{array}$$

14. 10
    3
    3,6,9-trimethyl-2,5,8-trioxadecane

15. $$\begin{array}{l} HO-CH_2-CH_2-O-CH_2 \\ \qquad\qquad\qquad\qquad\quad | \\ HO-CH_2-CH_2-O-CH_2 \end{array}$$

16. 3,5,8-trioxa-1-nonene

17. 8
    4
    sila
    2,2,7,7-tetramethyl-4-phenyl-3,6-dioxa-2,7-disilaoctane

CHAPTER 8 SUBSTITUTION PRODUCTS FROM AROMATIC HYDROCARBONS

1. ethylbenzene
   bromobenzene
   nitrobenzene
   nitrosobenzene

2. 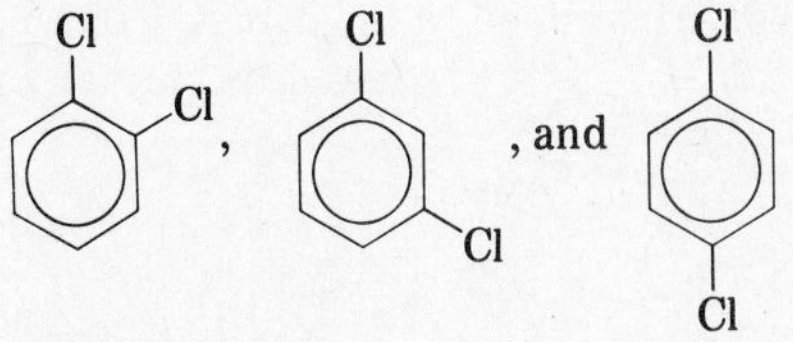 Cl, Cl ; Cl, Cl , and Cl, Cl

   o-dichlorobenzene, m-dichlorobenzene, and p-dichlorobenzene
   1,2-dichlorobenzene, 1,3-dichlorobenzene, and 1,4-dichlorobenzene

3. 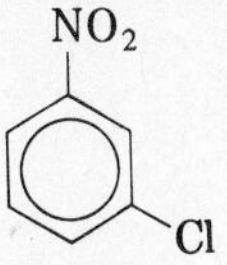 NO_2, Cl

   1-chloro-3-nitrobenzene

4. 1, 2, 3, 5
   2-chloro-1,3,5-trinitrobenzene

5. 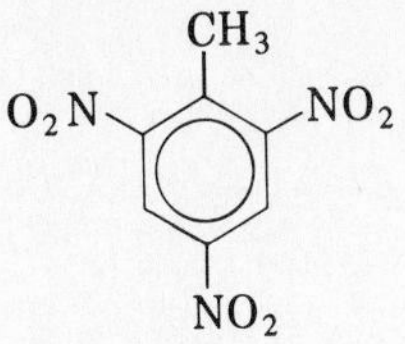 CH_3, O_2N, NO_2, NO_2

6. the use of benzene as the basis of a name requires the use of the smallest possible numbers, while the use of toluene as the basis of the name requires that the methyl group be on position number 1.
   2-methyl-1,3,5-trinitrobenzene

7. 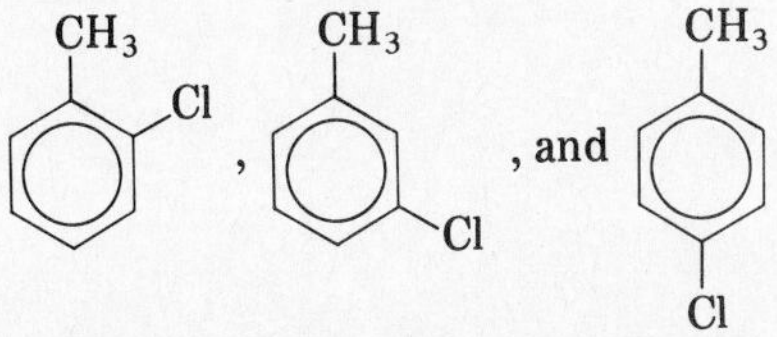 CH_3, Cl ; CH_3, Cl , and CH_3, Cl

   o-chlorotoluene, m-chlorotoluene, and p-chlorotoluene (or 2-chlorotoluene, 3-chlorotoluene, and 4-chlorotoluene)

8. 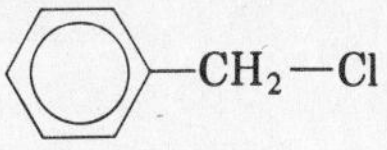 —CH_2—Cl

9. chloromethyl
   (chloromethyl)benzene

10. 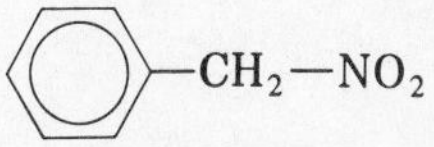 —CH_2—NO_2

    (nitromethyl)benzene
    without them, chloromethyl or nitromethyl might be mistakenly taken to indicate two substituents on the benzene ring

11. benzyl chloride
    benzyl iodide
    dibenzyl ether

12. 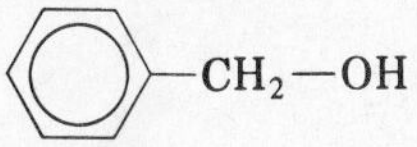 —CH_2—OH

    primary

13. 1-chloromethyl-4-nitrobenzene
    (2-methoxyphenyl)methanol

14. 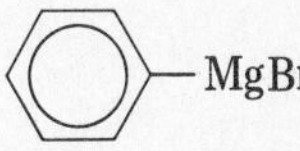 —MgBr or ϕ—MgBr

    or C_6H_5—MgBr or Ph—MgBr

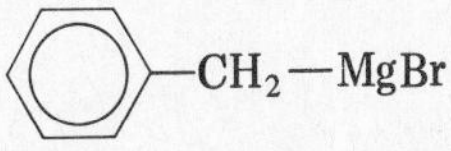

    —CH_2—MgBr or ϕ—CH_2—MgBr

    or C_6H_5—CH_2—MgBr or Ph—CH_2—MgBr

15. 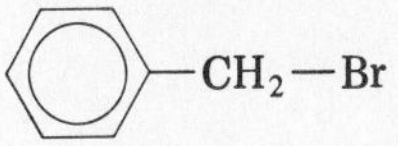 —CH_2—Br

    Br— —CH_3

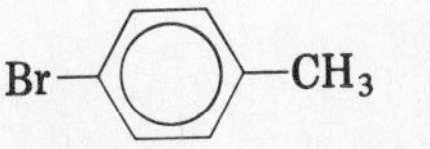

16. 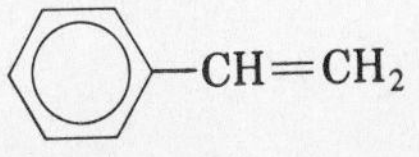 —CH=CH_2

17. 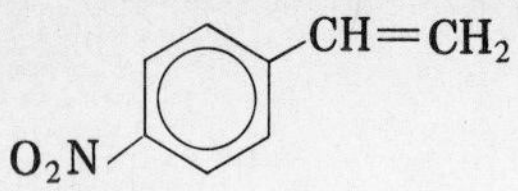 $CH{=}CH_2$

O_2N

1-ethenyl-4-nitrobenzene

$CH{=}CH_2$

18. 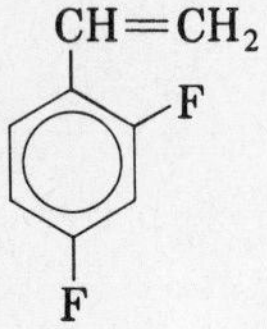 F

F

19. benzyl
4-benzyl-3-fluorostyrene

20.

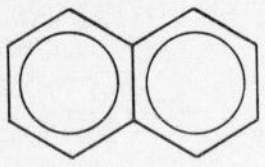

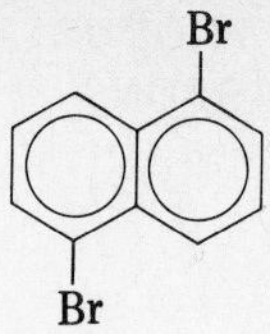

Br

Br

21. 1-methyl-2,4-dinitronaphthalene

22. hydrogens added to the parent, naphthalene
5-ethoxy-1,2,3,4-tetrahydronaphthalene

23.

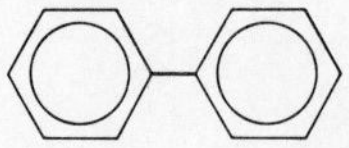

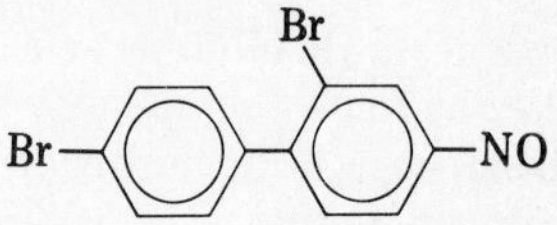

Br

Br— —NO

CHAPTER 9 ACIDS

1. $CH_3{-}CH_2{-}CH_2{-}COOH$
   $CH_3{-}CH{=}CH{-}COOH$
2. 6
   hex
   methyl
   3
   3-methylhexanoic acid
3. 9-decenoic acid
4. 7
   3 and 5
   3-benzyl-5-methyl-3-nitroheptanoic acid
5. $HC{\equiv}C{-}$
   $CH_3{-}CH_2{-}O{-}$
   $CH_2{-}CH_2{-}CH_2{-}CH{-}COOH$ (with $O{-}CH_2{-}CH_3$ on the first CH_2 and $C{\equiv}CH$ on the CH)
6. naphthalene
   naphthyl–O–
   naphthyl–$O{-}CH_2{-}COOH$
7. 5-phenyl-4-pentynoic acid
8. hexanoic acid
   ethoxy and 2,4-dibromophenyl
   4-(2,4-dibromophenyl)-3-ethoxyhexanoic acid
9. alkenoic
   alkynoic
10. the extension of the parent chain from the alkene linkage
    CH_3, H / C=C \ H, $CH{-}COOH$ ($CH_2{-}CH_3$ on the CH)
    CH_3, H / C=C \ H, $CH{-}COOH$ ($CH_2{-}CH_3$ on the CH)
11. 6
    $HOOC{-}CH_2{-}CH_2{-}CH_2{-}CH_2{-}COOH$
12. 2,3-dibromo-2,3-diphenylbutanedioic acid
13. E
    2-butenedioic acid
    (E)-2-chloro-3-methyl-2-butenedioic acid
    (locant for C=C not required, because it cannot be anything other than 2 in this compound)
14. H, COOH \ C=C / CH_3; $CH_2{-}C$; $CH_2{-}CH_2{-}CH{-}COOH$; CH_3 on the CH
15. $CH_3{-}C(CH_3)_2$–cyclohexane–COOH
16. 1-cyclooctenecarboxylic acid
    4-cyclooctenecarboxylic acid
17. 9
    benzene ring with COOH, COOH, COOH
18. 2,7-naphthalenedicarboxylic acid
    6-nitro-2,2′-biphenyldicarboxylic acid
19. dihydronaphthalene ring with COOH
20. 7
    $(HOOC)_2CH{-}CH_2{-}CH(COOH)_2$
21. 1,2,4,5-pentanetetracarboxylic acid

22. 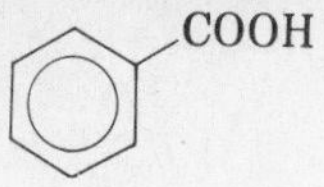 COOH

COOH

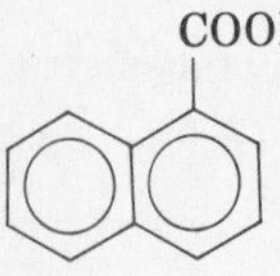

and

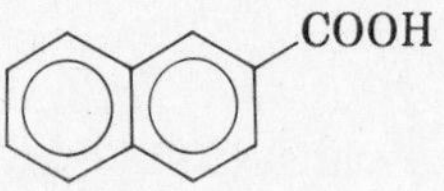

COOH

23. 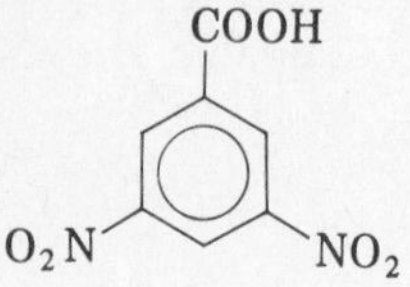 COOH

O_2N NO_2

24. 2
6-nitro-2-naphthalenecarboxylic acid
6-nitro-2-naphthoic acid

25. benzoic acid
1-propynyl
4-(1-propynyl)benzoic acid

26. prop
propion
but
butyr

27. CH_3—COOH
acetic acid

28. 2

29. chloroacetic acid
trichloroacetic acid

30. 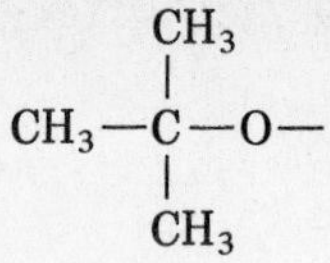
$$CH_3-\overset{\overset{\large CH_3}{|}}{\underset{\underset{\large CH_3}{|}}{C}}-O-$$

$$CH_3-\overset{\overset{\large CH_3}{|}}{\underset{\underset{\large CH_3}{|}}{C}}-O-CH_2-COOH$$

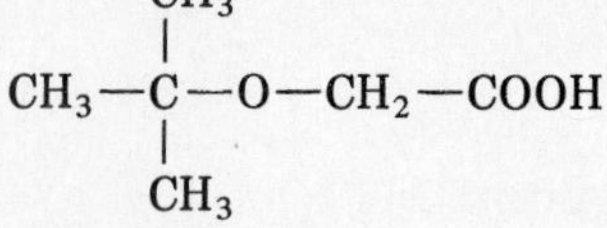

31. H—COOH
formic acid

32. peroxyformic acid <u>or</u> peroxymethanoic acid

33. 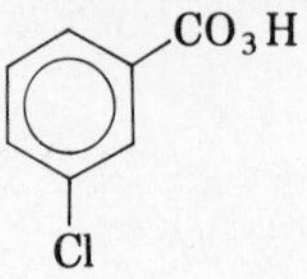 CO_3H

Cl

34. 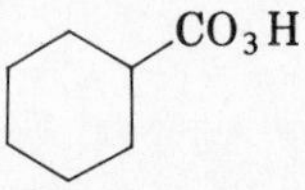 CO_3H

35. trifluoroperoxyacetic acid

36. $C_6H_5-SO_3H$

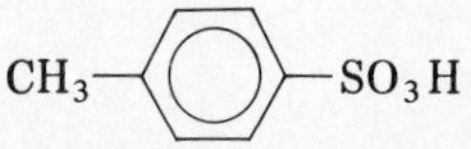

CH_3— —SO_3H

37. naphthalene
2
5-nitro-2-naphthalenesulfonic acid

38. trifluoromethanesulfonic acid

39. 3,3-dimethylcyclobutanesulfonic acid

CHAPTER 10 ACID DERIVATIVES

1. acetic acid
   acetic anhydride

2. $C_6H_5—COOH$
   $C_6H_5—CO—O—CO—C_6H_5$

3. $CF_3—CO—O—CO—CF_3$

4. cyclohexanecarboxylic acid
   cyclohexanecarboxylic anhydride

5. trifluoromethanesulfonic acid
   trifluoromethanesulfonic anhydride

6. (benzene ring with two adjacent COOH groups)
   COOH
   COOH

   (benzene ring fused to anhydride ring)
   C=O
   O
   C=O

7. $CH_2—C(=O)$
   | O
   $CH_2—C(=O)$

8. 1,8-naphthalenedicarboxylic acid
   1,8-naphthalenedicarboxylic anhydride

9. benzoic acid and acetic acid
   acetic benzoic anhydride (alphabetical)

10. butyric p-toluenesulfonic anhydride
    or butanoic 4-methylbenzenesulfonic anhydride

11. benzoic acid
    benzoyl chloride

12. $CH_3—COOH$
    $CH_3—CO—Cl$

13. 2-bromobutyric acid (or 2-bromobutanoic acid)
    2-bromobutyryl bromide (or 2-bromobutanoyl bromide)

14. $O_2N—C_6H_4—SO_2—Cl$ (para)

15. cyclopropanecarbonyl chloride

16. 2-naphthoyl chloride
    2-naphthalenecarbonyl chloride

17. formic acid
    methanoate
    formate

18. propanoic acid
    propionate
    propanoate

19. butyrate

20. $CH_3—CH_2—CH_2—CO—OCH_3$

21. sec-butyl
    cyclopentanecarboxylic acid
    cyclopentanecarboxylate
    sec-butyl cyclopentanecarboxylate

22. $CF_3—COOH$
    $CH_3—CH(CH_3)—CH_2—$
    $CF_3—COO—CH_2—CH(CH_3)—CH_3$

23. $CH_3—CH(CH_3)—COOH$
    $CH_3—CH(CH_3)—COO—CH_2—CH(CH_3)—CH_3$

24. allyl formate

25. $C_6H_5—O—C(=O)—CH_2—CH_3$
    $C_6H_5—CH_2—O—C(=O)—CH_3$

26. right
5
2,4
methoxy, methyl
2-methoxy-4-methylpentyl
7
5
3
5-chloro-3-methyl-5-heptenoic acid
5-chloro-3-methyl-5-heptenoate
2-methoxy-4-methylpentyl
5-chloro-3-methyl-5-heptenoate

27. 3,5-dinitrobenzoate
2-chloro-4-methylcyclohexyl
2-chloro-4-methylcyclohexyl
3,5-dinitrobenzoate

28. 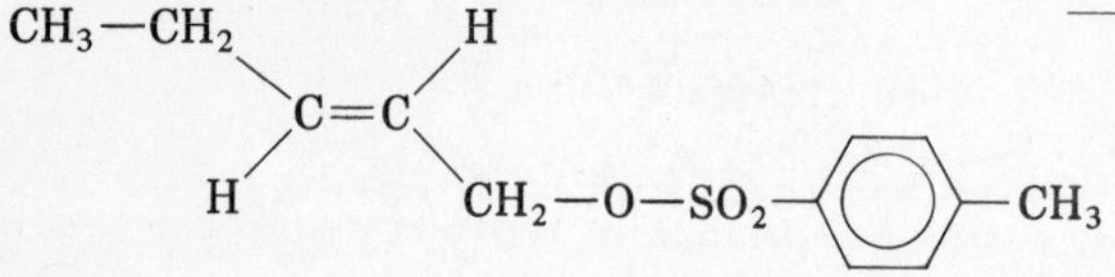 $CH_3—CH_2$ H
C=C
H $CH_2—O—SO_2—C_6H_4—CH_3$

29. 3-cyclooctenyl formate

30. $(CH_3—CH_2—CH_2—CH_2—O)_3B$

31. benzenesulfonic acid
benzenesulfonate
4-tert-butylcyclohexyl benzenesulfonate

32. $CH_2{=}CH—O—SO_2—CF_3$

33. sulfuric
$CH_3—(CH_2)_{10}—CH_2—O—SO_3^-Na^+$

34. 3,6-dioxaheptyl
3,6-dioxaheptyl acetate

35. ethoxycarbonyl
4-ethoxycarbonylcyclohexanecarboxylic acid

36. 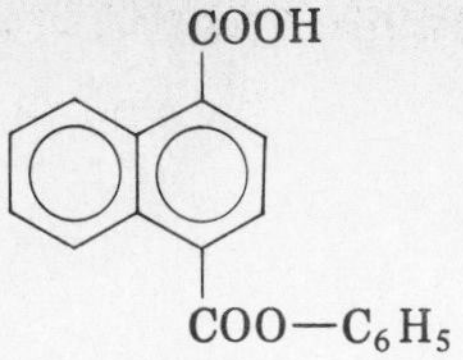 COOH
$COO—C_6H_5$

37. formic acid
formamide

38. 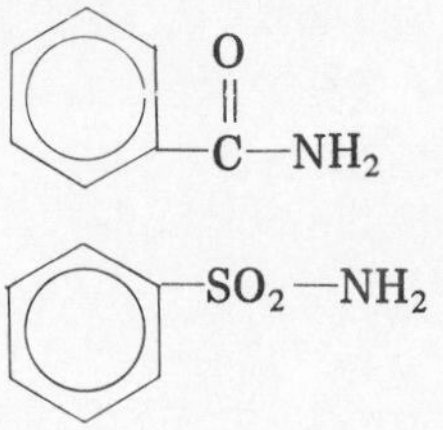 $C_6H_5—C(=O)—NH_2$

$C_6H_5—SO_2—NH_2$

---

39. 4-methylcyclohexanecarboxylic acid
4-methylcyclohexanecarboxamide

40. $CH_3—CH_2—CH(CH_3)—CH_2—CH_2—C(=O)—NH_2$

$CH_3—CH_2—CH_2—CH_2—CH_2—C(=O)—NH—CH_3$

41. $H—C(=O)—N(CH_3)—CH_3$

42. N-butylbenzenesulfonamide

43. benzamide
N and 3
N,3-dibromobenzamide

CHAPTER 11 ALDEHYDES AND KETONES

1. a ketone
2. an aldehyde
3. an aldehyde
   a ketone
4. pentane
   pentanal
5. 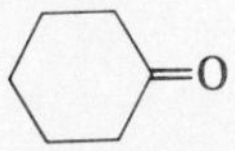 =O
6. $CH_3—CH_2—CHO$
7. $CH_3—CH_2—CH_2—CO—CH_3$
   $CH_3—CH_2—CO—CH_2—CH_3$
   2-pentanone
   3-pentanone
8. $CH_3—CH_2—CH—CO—CH_3$
   (Cl on CH)
   $CH_3—CH_2—CH—CH_2—CHO$
   (Cl on CH)
9. 3,3-dimethyl-2-butanone (Locant 2 not necessary but generally used.)
10. 5-phenyl-2-hexene
    5-phenyl-2-hexenal
11. 2-phenyl-2-heptene
    1
    5
    6
    6-phenyl-5-heptenal
12. 2-heptanone
    3,7-dimethyl-6-octenal
13. 2-cyclohexenone
    isopropenyl (or 1-methylethenyl)
    methyl
    2-methyl-5-(1-methylethenyl)-2-cyclohexenone or 5-isopropenyl-2-methyl-2-cyclohexenone
14. $CH_2=C=O$
15. 9
    2,6,6-trimethyl-1-cyclohexenyl
    3, 7
    trans-2-trans-4-trans-6-trans-8-nonatetraenal
    3,7-dimethyl-9-(2,6,6-trimethyl-1-cyclohexenyl)-trans-2-trans-4-trans-6-trans-8-nonatetraenal
16. 1
    cyclohexyl
    1-cyclohexyl-2-pentanone
17. $C_6H_5—CO—CH_2—CH_2—CH—CH_2—CH_3$
    (C_6H_5 on CH)
18. 1-cyclobutyl-2,3-dimethyl-1-butanone
19. $OHC—CH_2—CH_2—CH_2—CH_2—CHO$
    $CH_3—CO—CH_2—CO—CH_2—CH_3$
20. 4-phenylheptanedial
21. cyclopentanecarbaldehyde
    3-methylcyclobutanecarbaldehyde
22. 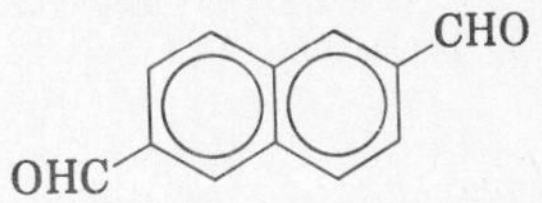 CHO
    OHC
23. 4-nitrobenzenecarbaldehyde
24. $CH_3—COOH$
    $CH_3—CHO$
25. isobutyric acid
    isobutyraldehyde
26. $C_6H_5—CHO$
    benzoic acid
    benzaldehyde
27. trichloroacetaldehyde
28. $CH_3—CH_2—CO—CH_3$
29. dicyclopropyl ketone
    diphenyl ketone
30. ethyl 2-naphthyl ketone
    1-(2-naphthyl)-1-propanone
31. $C_6H_5—CH_2—CO—CH—CH_2—CH_3$
    (CH_3 on CH)
    3-methyl-1-phenyl-2-pentanone

32. 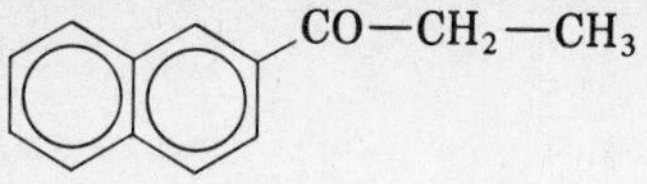 $CO{-}CH_2{-}CH_3$

33. benzophenone

34. 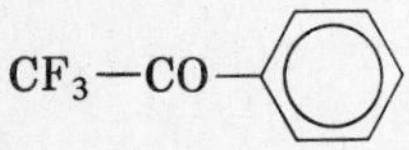 $CF_3{-}CO{-}$

35. 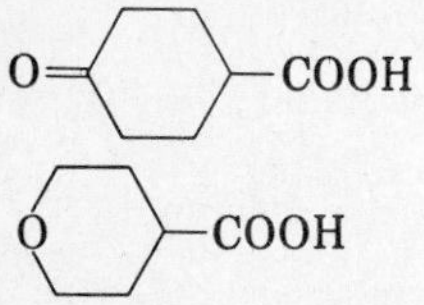 $O{=}$ $-COOH$

O $-COOH$

36. $H{-}\underset{\underset{\displaystyle O}{\|}}{C}{-}CH_2{-}CH_2{-}CH_2{-}CH_2{-}COOH$

an aldehyde

37. $CH_3{-}\overset{\overset{\displaystyle O}{\|}}{C}{-}CH_2{-}CH_2{-}CH_2{-}CH_2{-}CH_2{-}\overset{\overset{\displaystyle H}{|}}{C}{=}\underset{\underset{\displaystyle H}{|}}{C}{-}COOH$

38. cyclodecenecarbaldehyde
2
7-oxo-2-cyclodecenecarbaldehyde

39. vinyl 3-formylcyclopentanecarboxylate or
ethenyl 3-formylcyclopentanecarboxylate

40. 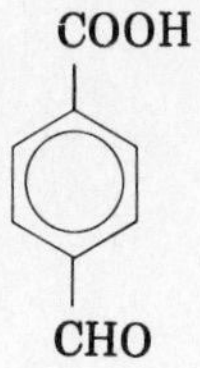 COOH

CHO

CHAPTER 12 AMINES AND RELATED CATIONS

1. secondary
2. tertiary
3. $R-NH_2$
   $R-NH-R$ or R_2NH
   $R-N(R)-R$ or R_3N
4. tertiary
   primary
5. secondary
   primary
6. primary
   primary
7. 2-butanamine
   sec-butylamine (or 1-methylpropylamine)
8. $CH_3-C(CH_3)_2-NH_2$ or $(CH_3)_3C-NH_2$
9. diphenylamine
10. $CH_3-CH_2-CH_2-CH_2-CH(CH_2-CH_3)-NH_2$
    3-heptanamine
11. 3-ethyl-3,5-dimethyl-1-hexanamine
12. 4-isopropyl-2-phenylcyclohexanamine
13. 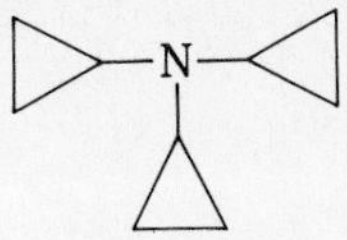 N
14. 2-methyl-1-propanamine (or isobutylamine)
    N-isopropyl-2-methyl-1-propanamine
15. benzene
    N,N-dimethylbenzenamine
16. 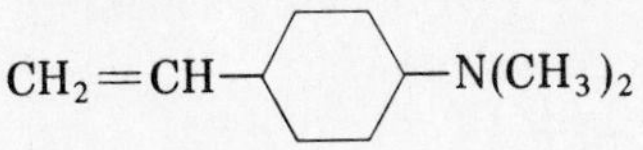 $CH_2{=}CH-C_6H_4-N(CH_3)_2$
    NH_3, amine
17. 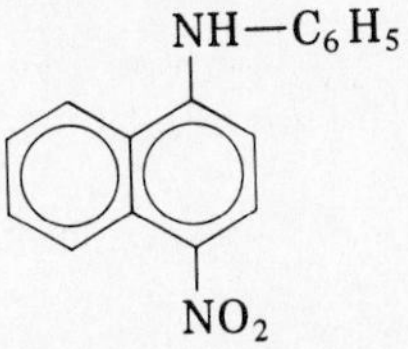 $NH-C_6H_5$
    NO_2
18. 2
    5-methyl-2-hexanamine
19. heptene
    N,N-dimethyl-6-phenoxy-4-heptene-1-amine
20. 6-methoxy-3-cyclooctenamine
21. N and 4
    N
    N-ethyl-N,4-dimethyl-2-hexen-2-amine
22. amino
    4-amino-3-hexanol
23. 4
    dimethylamino
    5
    5-dimethylamino-4-methyl-2-pentanone

---

24. $CH_3-CH_2-CH_2-CH_2-CH_2-CH_2-CH_2-CH(O-CH_2-CH_3)-CH(N(CH_2-CH_3)_2)-C(=O)-O-CH_2-CH_3$

---

25. 1,4-butanediamine
    1,5-pentanediamine

26. $H_2N-CH_2-CH_2-CH_2-CH_2-CH_2-CH_2-NH_2$

27. $$CH_3-\underset{\displaystyle NH_2}{\underset{|}{CH}}-\overset{\displaystyle CH_3}{\overset{|}{CH}}-\underset{\displaystyle NH_2}{\underset{|}{CH}}-CH_3$$

28. 1,2,5-pentanetriamine

29. $(CH_3)_2N-CH_2-CH_2-N(CH_3)_2$
$H_2N-CH_2-CH_2-NH_2$

30. diethylamine (*or* *N*-ethylethanamine)
diethylaminium chloride (*or*
N-ethylethanaminium chloride)

31. (2-naphthyl) $\overset{+}{N}H_3\ Cl\bar{O}_4$

32. 4-*tert*-butyl-*N*,*N*-dipropylcyclohexanaminium ion

33. $HO-\overset{+}{N}H_3\ Cl$
hydroxylaminium chloride

34. $$CH_3-\underset{\displaystyle +NH_3}{\underset{|}{CH}}-CO\bar{O}$$

35. $$CH_3-\underset{\displaystyle CH_3}{\underset{|}{CH}}-CH_2-\underset{\displaystyle +NH_3}{\underset{|}{CH}}-CO\bar{O}$$

36. 2-aminio-3-hydroxybutyrate
(*or* 2-aminio-3-hydroxybutanoate)

37. $C_6H_5-CH_2-\overset{+}{N}(CH_3)_3\ \bar{Cl}$

38. *N*-ethyl-*N*-octyl-1-octanamine
diethyldioctylammonium iodide

39. $R_4\overset{+}{N}\ \overset{-}{O}H$

40. (2-hydroxyethyl)trimethylammonium hydroxide

41. trimethylammonioacetate (*or*
trimethylammonioethanoate)

1. 8
   bicyclooctane
2. bicyclopentane
3. 3, 2, 1
4. 2, 1, 0
5. bicyclo[2.1.0]pentane
6. bicyclo[4.2.0]octane
   bicyclo[4.1.1]octane
   bicyclo[3.3.0]octane
   bicyclo[5.1.0]octane
   bicyclo[2.2.2]octane
7. 2
   7
8. 7
   3
9. Cl
   Cl
10. CH_3 — CH_3
11. bicyclo[2.2.1]hept-2-ene
12. CH_3 CH_3 CH_3
13. 3,7,7-trimethylbicyclo[4.1.0]hept-3-ene
14. bicyclo[2.1.1]hexan-5-ol
15. 1,7,7-trimethylbicyclo[2.2.1]heptan-2-one
16. COOH
17. O
18. 9-oxabicyclo[4.2.1]nonane
19. bicyclo[3.2.2]nonane
    aza
    3-azabicyclo[3.2.2]nonane
20. 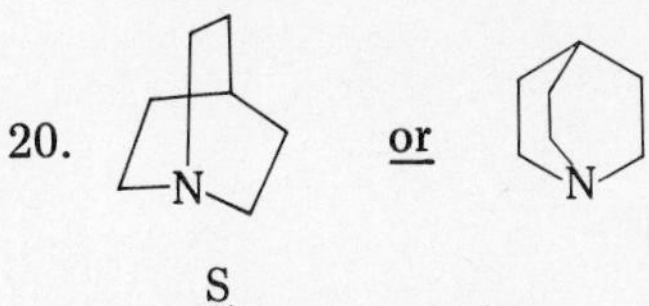

    S
    or
    S
21. 3
    8-azabicyclo[3.2.1]octane-2-carboxylic acid
    methyl 3-benzoyloxy-8-methyl-8-azabicyclo[3.2.1]octane-2-carboxylate
22. CH_3 CH_3
    CH_3
23. 2
    2-nitronorbornane
24. Cl
25. 2,5-norbornadiene
26. OH
    7-norbornanol
27. O
    2-norbornanone

28.

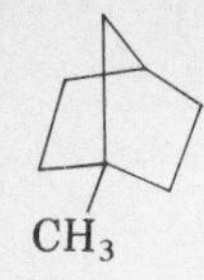

29.

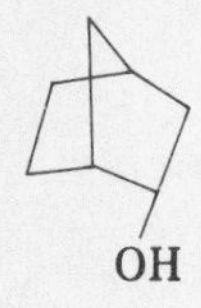

30. endo-2-bromonorbornane

31. exo-2-norbornanamine (or 2-norbornylamine)

32.

$O-\overset{\overset{\displaystyle O}{\|}}{C}-CH_3$

$O-SO_2-C_6H_5$

CHAPTER 14 NOMENCLATURE OF REACTION INTERMEDIATES

1. $CH_3-\underset{\cdot}{\overset{CH_3}{\overset{|}{C}}}-CH_3$
   tertiary
2. $C_6H_5-\dot{C}H_2$
   primary
3. propyl radical
   1-ethylpropyl radical
4. triphenylmethyl radical
5. 3-chlorobenzoyl (or m-chlorobenzoyl or 3-chlorobenzenecarbonyl) radical
   4-chlorophenoxyl (or p-chlorophenoxyl) radical
6. $C_6H_5-CO-O\cdot$
   $(CH_3)_2N\cdot$
7. 4-tert-butylcyclohexyl radical
   secondary
8. methyl-2-naphthylaminyl (or N methyl-2-naphthalenaminyl) radical
9. $Cl_2\ddot{C}$
   $C_6H_5-\ddot{C}H$
10. 6
    7
11. radical is a functional class name, not the name of a compound, but carbene is a parent compound name which is modified by prefixes that are part of the same word
12. propylcarbene (or propylmethylene)
    butyl radical
13. $H-C\equiv C-\ddot{C}H$
    $NC-\ddot{C}-CN$
14. 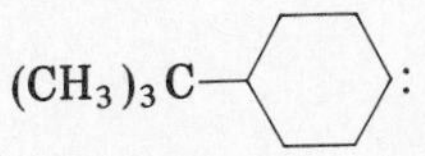 $(CH_3)_3C-$(cyclohexane ring)$:$
    4-tert-butylcyclohexylidene
15. 2-vinyl-1-carbenacyclopropane
    2-vinylcyclopropylidene
16. 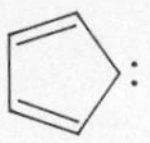 (cyclopentadiene ring)$:$
    1-carbena-2,4-cyclopentadiene
17. $C_6H_5-\ddot{N}:$
    $NC-\ddot{N}:$
18. methoxycarboxyl
    methoxycarbonylnitrene
19. 2-methylbutyrylnitrene (or 2-methylbutanoylnitrene)
20. 1-ethylpropyl cation
    secondary
    3-pentyl is not a correct name for any alkyl group
21. 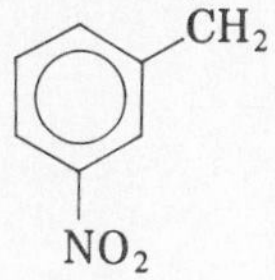 (benzene ring with $\bar{C}H_2$ and NO_2)
    primary
22. $CH_2=CH-\overset{+}{C}H_2$
    $CH_2=\overset{+}{C}H$
23. 1,6-dimethylcyclodecyl cation
    tertiary
24. 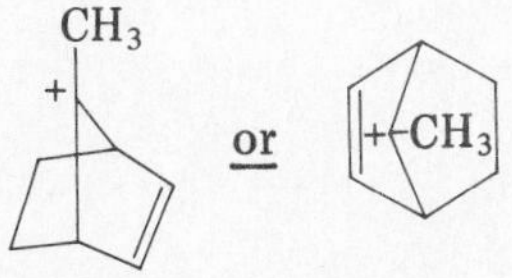 (CH_3, +) or (+CH_3)
25. 1,1,2-triphenylethyl anion
    tertiary
26. A: 1,3,5-cycloheptatriene
    B: 2,4,6-cycloheptatrienyl radical
    C: 2,4,6-cycloheptatrienylidene or 1-carbena-2,4,6-cycloheptatriene
    D: 2,4,6-cycloheptatrienyl cation
27. trichloromethyl anion
    methoxycarbonyl cation